Exploratory and explanatory statistical analysis of spatial data

CORNELIS P. A. BARTELS and RONALD H. KETELLAPPER, editors
University of Groningen

Martinus Nijhoff Publishing
Boston/The Hague/London 1979

Distributors for North America:
Martinus Nijhoff Publishing
Kluwer Boston, Inc.
160 Old Derby Street
Hingham, Massachusetts 02043

Distributors outside North America:
Kluwer Academic Publishers Group
Distribution Centre
P.O. Box 322
3300 AH Dordrecht, The Netherlands

Library of Congress Cataloging in Publication Data

Regional Science Symposium, University of Groningen, 1977.
Exploratory and explanatory statistical analysis of spatial data.

Contains, for the most part, papers presented at a Regional Science Symposium, held at the Faculty of Economics of the University of Groningen in the Netherlands.

Includes bibliographies.

1. Regional economics—Statistical methods—Congresses. 2. Regional planning—Statistical methods—Congresses. I. Bartels, Cornelis, P.A.
II. Ketellapper, Ronald H. III. Title.
HT391.R3374 1977 330'.01'82 79-13142
ISBN 0-89838-004-9

Printed in the United States of America.

Exploratory and explanatory statistical analysis of spatial data

Preface

In September 1977 a "Regional Science Symposium" was held at the Faculty of Economics of the University of Goningen in the Netherlands.

The impetus in organizing this symposium was the recent establishment at the Faculty of Economics of a group engaged in teaching and research within the field of regional science. The aim of the symposium was to familiarize university members with regional science and to introduce the new group to both the national and international scene. Two separate topics of potential interest to both researchers and policy-makers were selected.

The first theme, spatial inequalities and regional development, was chosen because of its central place in regional science. Authors from several disciplines were asked to approach this theme from a general, policy-oriented point of view. This ensured the spotlighting of the various dimensions of spatial inequality and its implications for regional policy. The results of their efforts have been collected in a *volume* entitled *Spatial Inequalities and Regional Development*.

The second theme focussed on spatial statistical analysis. This branch of statistics is a relatively new one. It is receiving growing attention from researchers in the field of applied regional science. The conference dealing with this topic concentrated on recent research results related to the use of appropriate statistical and econometric methods for analyzing spatial data. The papers concerned have been collected in another *volume*, entitled *Exploratory and Explanatory Statistical Analysis of Spatial Data*.

Both volumes contain, for the most part, papers presented at the symposium. Some additional papers have been included to improve the consistency of the volumes. All contributions have been revised prior to final publication. In this process critical comments made by participants at the symposium have proven to be very helpful. We believe that these efforts have helped considerably to improve the quality of both volumes.

Groningen, April 1978 — The editors

Contents

List of contributors

Cornelis P.A. Bartels is assistant professor in regional economics at the University of Groningen (Netherlands). His publications include a book on economic aspects of regional welfare and several articles on development economics, income distribution, regional unemployment and econometric techniques applied to spatial data.

Robert J. Bennett is affiliated with the Department of Geography of the University of Cambridge. His various publications include books on spatial time series and environmental systems, and articles on the identification, representation and estimation of dynamic spatial models, optimal control models of regional economies, and techniques for non-stationary parameter estimation.

Andries S. Brandsma was student in the Faculty of Econometrics at the University of Groningen, Netherlands. His master thesis focused on methods to account for spatial autocorrelation in the estimation of regional economic models.

Leslie W. Hepple is at the Department of Geography of the University of Bristol, England. He published studies on several aspects of econometric estimation with spatial data, e.g., the use of stochastic process theory in spatial analysis, spectral analysis, maximum likelihood estimation and Bayesian analysis of regional models.

Ronald H. Ketellapper is assistant professor in econometrics at the University of Groningen (Netherlands). His research is mainly in the field of the estimation of econometric models, the analysis of errors-in-variables models, and the study of techniques to test for spatial autocorrelation.

Stan Openshaw is lecturer at the Department of Town and Country Planning of the University of Newcastle upon Tyne, England. His research concentrated on several aspects of the use of spatial interaction models in planning, especially the specification of deterrence functions, the determination of optimal zonings, and properties of alternative specifications of such interaction models.

Brian D. Ripley is associated with the Department of Mathematics of the Imperial College in London. He has published articles on the theoretical and empirical analysis of stationary point process in space.

Folke Snickars is member of the research group for Urban and Regional Planning at the Royal Institute of Technology in Stockholm. His research includes studies on the application of information theory in regional science, and the analysis of regional migration.

Bernd Streitberg works in the field of econometrics and statistics at the Freie Universität, Berlin. His research concentrated on the estimation of linear multivariate models with dependent data.

K.C. Tan worked at the Department of Geography of the University College, London. He published on optimal control theory for linear econometric systems with linear equality and inequality constraints on the control variables.

Jan van Est is research fellow at TNO, Delft, Netherlands. He works on the estimation and application of large scale spatial distribution models.

Jan van Setten is research fellow at TNO, Delft, Netherlands. His research includes the estimation of spatial interaction models; and the specification of mathematical models of spatial patterns in the service system.

Part 1: Introduction

1. General introduction

In this volume a number of papers dealing with the statistical analysis of spatial data have been collected. A subdivision has been made between papers of an exploratory nature and those which deal with aspects of explanatory analysis.

In his survey article, Cornelis Bartels discusses several statistical methods that have been used for the exploratory and explanatory statistical analysis of spatial data. The different techniques were selected because of their operational value in empirical applications.

The following three papers have in common their aim to give a descriptive, exploratory analysis of spatial data. Brian Ripley reviews several methods to detect regularity and clustering in spatial data which are in the form of maps. Evaluation of these methods leads to the conclusion that traditional methods, which summarise the data in a few statistics, do not perform satisfactorily. More advanced techniques, based on explicit stochastic models and employing functional statistics should be used. Some new statistics are defined and applied to several sets of data. Folke Snickars tackles the problem of how to construct an interregional input-output table if incomplete information is available. Starting from incomplete survey data, a method is described which generates an input-output table which is consistent with non-survey macro data. The method is applied to Swedish data. Andries Brandsma and Ronald Ketellapper evaluate several procedures for testing spatial autocorrelation among regression disturbances. The Moran statistic, using alternative residuals, and a likelihood ratio test are included in their Monte Carlo experiments. It appears that ordinary least-squares residuals, inserted in the Moran statistic, yield the most attractive testing procedure.

A following set of four papers relates to topics from explanatory statistical analysis, using spatial data. Bernd Streitberg presents a rather general theory for treating dependencies among spatial data in a causal explanatory model. It is based on a decomposition of the covariance matrix of the observed variable. Such a decomposition is attractive when dealing with heteroscedasticity, spatial stationary stochastic processes, aggregation in space, transaction tables and spatial moving average processes. Operational methods to estimate regression parameters in case the observations on the dependent variable follow such a covariance structure are also indicated. Leslie Hepple considers a spatial linear regression model, with the disturbances generated by a spatial autoregressive process. It is then indicated how a Bayesian approach may be employed for purposes of parameter estimation. It is argued that examination of the marginal and conditional posterior distributions may facilitate the analysis of robustness of the model to specification errors. The approach is applied to spatial data, which previously have been analysed by means of maximum likelihood procedures. Stan Openshaw focuses on the selection of spatial interaction models. He pleads for evaluation of alternatively estimated and specified models on the basis of their forecasting performance, rather than the model's goodness-of-fit. The argument is illustrated for data of a regional system, which have been collected for two distinct years. Effects of different estimation methods, errors in variables and different levels of aggregation also receive attention. Jan van Est and Jan van Setten develop a linear least squares estimation method for a singly-constrained spatial interaction model. This estimation procedure is compared with a traditional maximum likelihood approach, both theoretically and empirically.

The final paper of the book questions how to use an estimated explanatory spatial model for purposes of regional policy. Ron Bennett and K. Tan start from the goal to formulate rules of stochastic control theory in a regional context. It appears that stochastics can be taken into account only in a partial way, using the traditional tools of control theory. The authors indicate how their control algorithm for dynamic simultaneous regional models with a quadratic objective function may be adapted when a set of linear constraints is imposed on the control variables. The methods have been illustrated by means of an example, linked to an empirical use of the approach for evaluating British regional policy.

It is the hope of the editors that the joint efforts of economists, geographers, statisticians and econometricians which contributed to this book will inspire the reader to explore further the numerous remaining problems in the field of spatial statistical analysis.

Cornelis P.A. Bartels
Ronald H. Ketellapper

Acknowledgement: thanks are due to Koos Spanjer for his help in correcting the proofs.

2. Operational statistical methods for analysing spatial data

CORNELIS P.A. BARTELS*

2.1 INTRODUCTION

In applied regional science a fast growth can be observed in the number of statistical methods that are used for analysing spatial data. These methods incorporate both well-developed standard a-spatial statistical methods, and new adaptations of such methods to the specific spatial context. The first set of techniques dominates and seems to have yielded less controversial and better interpretable results than the second. However, for both approaches it is of prime importance to give a critical evaluation of their appropriateness for application to a set of spatial data and their ability to extend the insight into empirical spatial phenomena. In many studies such an evaluation is virtually absent, at least in a general context.

In this paper we intend to give a non-exhaustive survey of different statistical methods that can be used to analyse spatial data. We shall try to evaluate their usefulness, using as the most important criterion the operationality of the methods in empirical research. The survey has to be non-exhaustive, since covering this whole fast developing area with contributions from rather different disciplines (geography, econometrics, statistics, economics, biology) would lead us far beyond the intention of an introductory paper like the present one. Besides, other partial surveys have appeared recently, viz. Berry (1971), Hepple (1974), Cliff and Ord (1975), Cliff et al. (1975). The exposition to be given below tries to focus on elements that have gotten no or little attention in these studies. Specifically, the following two principles have been chosen for the composition of this contribution.

*Faculty of Economics, University of Groningen, P.O. Box 800, Groningen, the Netherlands.

First, since we shall employ at different places our experiences gathered during earlier research, most attention is given to the analysis of economic data with a spatial dimension. Secondly, it is explicitly recognized that the ultimate purpose of spatial statistical analysis is to obtain a quantitative, explanatory analysis of spatial phenomena. This allows for tests on hypothesized causal relationships, which are supposed to describe temporal and spatial processes. However, there are some reasons which may prevent such a parametric formulation of causal relations. As commonly encountered deficiencies in applied spatial research we mention:

- the lack of clear operational theories of spatial phenomena and interactions. This may be partly explained by the fact that regional science is a relatively new discipline.
- the lack of appropriate data, since the importance of collecting suitable spatial data-sets has long waited admission, and it is in many cases not yet recognized.

If such deficiencies make a straightforward parametric causal analysis impossible, it is quite natural to resort to the use of non-parametric, exploratory statistical techniques to discover peculiarities of the data at hand, like (ir)regularities in the series and associations between different data. So we shall discuss both the exploratory and explanatory statistical analysis of spatial data. Several methods of analysis will be reviewed, and we shall try to indicate their merits for operational use. Before we start this discussion, we shall briefly consider the type of data that forms the point of departure of this paper. Next, three sections (Sections 2.3, 2.4, 2.5) will be devoted to various exploratory statistical methods. In Sections 2.6 and 2.7 statistical procedures of an explanatory character receive our attention. Section 2.8 treats some remaining topics.

2.2 THE STRUCTURE OF SPATIAL DATA

Data on variables defined for units in space may be available in several forms. The most general case is where we have variables defined on both the time- and space-domain. Let us denote the time-domain with the linear index-set T, and an element of this set by $t \in T$. If spatial stock variables are studied, the space-domain has two dimensions. The spatial co-ordinates are denoted by $r_1, r_2 \in R^2$, where R is the set of real numbers. In many cases it will be required that these co-ordinates are equally spaced, to make application of theoretical statistical concepts possible. A variable X defined on T and R^2 is hence denoted by $X(r_1, r_2, t)$ and its observation by $x(r_1, r_2, t)$. If one studies spatial flow-variables, more indices are required to denote the variables. For example, the flow from unit (r_1, r_2) to unit (r_3, r_4) in time period t is denoted by $X(r_1, r_2, r_3, r_4, t)$.

This general definition of spatial variables is not employable and/or desirable in many studies in this field. For example, if one is concerned with socio-economic variables in a spatial context, one will frequently restrict the space-domain to one dimension if the exact location of each spatial unit is considered as not important. In such a situation we may represent the spatial variable by $X(r, t)$ with $r \in S$ and S a linear set of spatial indices. If observations on this variable are available, we have a time-series of cross-sectional data at our disposal. For many spatial data such time-series are absent, and we only have a cross-section of data for different points in space. In this case the variable being observed is $X(r)$. If flows between points in space are being studied, the notation will be $X(r, r')$ with $r, r' \in S$. In the sequel we shall make use of this notation.

2.3 METHODS BASED ON SIMPLE CORRELATIONS BETWEEN CROSS-REGIONAL DATA.

A number of statistical techniques is available for an exploratory analysis of spatial patterns. We distinguish methods based on simple correlations between spatial data (this section) and methods derived from time-series analysis (Sections 2.4. and 2.5).

There are yet some other methods designed for the analysis of spatial patterns. For example, one may want to test whether an observed spatial pattern is generated by a random distribution of the variable over space. Several test procedures have been designed for this purpose. If such tests indicate nonrandomness, a next step involves the specification of some alternative model, e.g. by means of a probability density function, which may have generated the observed phenomenon. For a more extensive discussion of such topics we refer to the contribution of Ripley in this volume. Other introductory references include Besag (1974), Cliff and Ord (1975) and Ripley (1977).

So in this section we discuss a first set of exploratory techniques, which all use simple coefficients of correlation between cross-regionally observed variables as the point of departure.

2.3.1 Multivariate correlation analysis

Suppose we have a set of regionally observed variables at one instant in time, of the type $X(r)$ discussed above. If we denote the number of regions by R and the number of variables by I the observations can be written as an R x I data-matrix. In many cases one is interested in knowing which pair-wise associations exist among the variables, as observed in the spatial system that is being studied. A quite natural first step is to consider Pearson's product-moment coefficients of correlation. For variables i and i' this coefficient of correlation is denoted by $\rho(i, i')$. It equals

$$\rho(i, i') = \mathrm{E}(X_i - \mu_i)(X_{i'} - \mu_{i'}) \,/\, (\sigma_i \, \sigma_{i'}) \tag{3.1}$$

with

$$\begin{aligned} \mu_i &= \mathrm{E}(X_i) \\ \sigma_i &= \{\mathrm{E}(X_i - \mu_i)^2\}^{1/2} \end{aligned} \tag{3.2}$$

An estimator of(3.1) may be obtained by substituting the sample moments; this yields $\hat{\rho}\,(i, i')$.

These coefficients can be written as elements of the I x I estimated correlation matrix $\hat{R}$, the elements of which express pair-wise linear associations among the spatially observed variables. However, this presentation is not very compact and does not shed light on common components present in several variables. In order to obtain a more compact representation of these linear associations additional statistical methods may be used. In the following sections we discuss two of them.

2.3.2 Factor analysis

A first statistical method that has much been used to analyse multivariate data is factor analysis. Factor analysis may be considered as a special case of the classical "errors in variables" structural models in econometrics. These "errors in variables" models take part of the independent variables as unobservable, while in factor analysis all independent variables are considered unobservable (cf. also Goldberger, 1972). It is assumed that a number of observed variables can all be expressed as a linear combination of some unobservable hypothetical variables (the common factors) and a disturbance (unique factor). This so-called common factor model may be written as

$$Z = AF + \epsilon \tag{3.3}$$

with

Z an I random vector of standardised, observable variables
F a K random vector of unobservable common factors
ϵ an I random vector of disturbances or unique factors
A and $I \times K$ coefficient matrix.

General additional assumptions are

- $E(F) = 0$, $E(\epsilon) = 0$
- $E(F\,\epsilon') = I$, i.e. no correlation between unique and common factors
- $E(\epsilon\,\epsilon') = I$; mutually uncorrelated, normalised unique factors
- $E(F\,F') = \Sigma_{ff}$; a diagonal matrix implying mutually uncorrelated common factors (in oblique factor analysis this restriction is not imposed)
- in principal factor analysis one requires $E(F\,F') = I$ (scaling)
- in principal components analysis one assumes $\epsilon = 0$ and $K = I$, i.e. the whole variance of the observable variables is attributed to the common factors.

Estimates of the coefficient matrix A, the matrix of coefficients of correlation between variables and factors (the factor structure matrix which contains important interpretable information), and the realisation of the factors may be obtained by imposing additional restrictions. If we derive each factor such as to account for the maximum of total variance of the variables after previous factors have been partialled from the variables, the calculation procedure amounts to the determination of eigenvalues and eigenvectors of the estimated correlation matrix $\hat{R}$ (which may be modified if $\epsilon \neq 0$, since then an estimate of the part of variance of z_i explained by the common factors is used as the i-th diagonal element). The eigenvalues indicate which part of the total variance of the variables is represented by the factor associated with each eigenvalue. Hence, the eigenvalues can be used to select a small number of factors representing most of the variance of the observed variables. Further details on the procedure of calculation and selection rules may be found in Dhrymes (1970) and Mulaik (1972).

The procedure briefly described above does not yield a unique solution. This flexibility allows us to apply some transformation in order to obtain a simpler parametric structure which makes interpretation of the unobservable factors easier. Hence, an ex-post redefinition of unobservable variables and parameters is allowed in this approach, which makes the procedure rather flexible.

In empirical applications this technique is mostly used to summarise the information as presented by a complete data-set into a small number of artificial variables, which contain nearly the same information. For its use with spatial data, one starts mostly from a set of variables X_1 (r), X_2 (r), ..., $X_I(r)$ observed for R regions. Two particular purposes may be mentioned, which make the use of factor analysis attractive.

1. In order to get a description of regional welfare and the economic structure of regions, many variables may be selected, each giving a partial description. Since all will be correlated to some extent, each variable does not represent a separate dimension. Therefore, one may prefer to summarise this information into a smaller number of artifical variables. This may facilitate the discovering of general patterns in the data. To guarantee that each artificial variable (or factor) represents a separate aspect of the data, one requires the factors to be uncorrelated. Having obtained a small number of uncorrelated factors, one may select some to obtain a ranking of regions (for example, if the first factor explains a large part of the variances the estimated values of this factor can be used as a criterion to obtain a rank-order), or to suggest some underlying explanatory variables. Examples of this approach are the study of economic structure by Jaumotte and Paelinck (1971) and the welfare studies by Adelman and Morris (1965), (1973), Smith (1973),

and Van Rompuy and Tejano (1975). The two latter studies use factor analysis to derive weights of a regional welfare function (i.e. the first common factor is used as an indicator of regional welfare). These weights result from some arbitrary statistical optimisation criteria, which makes it impossible to associate them with trade-offs between welfare-elements as might exist in reality. Interpretations of the results seem therefore of limited value, and a better approach could perhaps use completely a priori specified welfare weights.

2. Considering the set of variables introduced above, one may already possess a hypothesis that suggests one of the variables as being related to the others. Since in many cases no clear-cut theory is available that suggests the functional form and the variables to be included, one may in principle select a large number of variables that may be expected to possess some relation with the dependent variable. However, such a large number of associated variables may cause problems for statistical estimation, e.g.,
   - the number of variables may become too large as compared with the number of observations
   - a high degree of multicollinearity may exist among the variables.

   In this situation one may use factor analysis to summarise the original set of variables in a smaller set of uncorrelated components. The most important components are then included in the regression analysis. This approach was applied to regional data by Geary (1954). A recent contribution is Hampton and Rayner (1977). A discussion of properties of the resulting parameters is contained in Chen and Iglarsh (1976), Johnson et al. (1973), McCallum (1970), and Scott (1966). Another possibility consists of using the results to select a subset of variables that express a large part of the variation in the data and are mutually only weakly correlated. This procedure was followed in a study of variables associated with indicators of regional income distributions, cf. Bartels and Bertens (1976). Starting with 14 variables observed for 39 regions, it appeared that 7 uncorrelated common factors accounted for 99.7% of the variability in the data. Each factor could be associated with an observable variable, so that these variables could be used in a subsequent regression analysis. (If a theoretical model incorporating all 14 variables could be hypothesized, using a subset of variables produces biased parameter estimates. If no such hypothesis is available, this drawback is less severe.) It seems that this procedure has some advance above stepwise regression analysis, since the results of the latter technique may depend on the order in which variables enter the analysis.

An alternative selection procedure, which seems also attractive for spatial data, is given by interdependence analysis, cf. Boyce et al. (1974). This technique aims at selecting a subset of J variables ($J < I$) in some optimal way. One chooses a subset of J variables, which are used as explanatory variables in $(I\text{-}J)$ regressions, where the remaining $(I\text{-}J)$ variables enter successively as dependent variables. For each regression the coefficient of determination R^2 is computed, and the minimum value of R^2 for this set of regressions is recorded: $R^2{}_{\min,k}$. The process is repeated for all possible subsets of J variables, i.e. $k=1, \ldots, \binom{I}{J}$

As the optimal subset one selects that subset which is associated with the maximum of $R^2{}_{\min,\,k}$. Cf. Also Nijkamp (1976) for a discussion. As a possible drawback of the method we mention the fact that one has to choose a priori the value of J while in factor analysis this is determined by the results of the statistical analysis. An advantage is that the original observed variables are retained during the analysis.

2.3.3 Canonical correlation analysis

In the preceding section we dealt with a set of regionally observed variables, for which we wanted to discover mutual associations. Canonical correlation analysis deals with two sets of variables and aims to investigate the associations between these sets of variables. For example if one studies a set of level of living indicators one might want to know which combination of these indicators has highest correlation with a combination of several economic variables. The linear combinations of both sets of variables that possess the highest mutual correlation may be found by canonical correlation analysis. Let $\boldsymbol{X}$ be an I-vector of observable random variables and $\boldsymbol{Y}$ a J-vector of observable random variables ($J \leqq I$), which are assumed to be standardised. One looks for the I-vector of weights $\boldsymbol{a}_1$ and the J-vector of weights $\boldsymbol{b}_1$ so that the linear combinations

$$\nu_1 = \boldsymbol{a}_1' \boldsymbol{X} \tag{3.4}$$

and

$$w_1 = \boldsymbol{b}'_1 \, \boldsymbol{Y} \tag{3.5}$$

possess a maximum mutual correlation. In order to derive a unique solution one has to impose an additional restriction on these linear combinations, i.e. their variances have to be 1. If $\boldsymbol{a}_1$ and $\boldsymbol{b}_1$ have been found, which requires an estimate of the correlation matrix $\boldsymbol{R}$ and solving a certain determinantal equation, one proceeds by deriving other linear combinations, or canonical variables, such that these new canonical variables are uncorrelated with the preceding ones. More information on the calculation procedure is given in Anderson (1958), Bartels and Nijkamp (1976) and Dhrymes (1970).

We employed this technique in a former study (Bartels, 1977a, Chapter 8) to investigate the associations between income distribution vectors for a set of regions and several socio-economic variables. The aim was to find optimal weights, which could be used to summarise the detailed information on intra-regional income distributions into a compact scalar measure of intra-regional income inequality. Other examples of application with spatial data are Hampton and Rayner (1977), Ranner (1974), and Van der Knaap and Lesuis (1976).

A possible difficulty in the calculations may arise if the data within each set appear to be very strongly correlated, since then the correlation matrices may become ill-conditioned (as indicated by a small determinant) and the results very sensitive to small calculation errors. To circumvent this difficulty it has been suggested to use ideas from ridge regression (cf. Marquardt and Snee, 1975, for an introduction of ridge regression). The procedure implies addition of constants to the diagnonal elements of the variance-covariance matrices, cf. Vinod (1976) for a fuller discussion.

In this section we considered some frequently used multivariate techniques, that have appeared useful for an exploratory analysis of a cross-section of spatial data. We now proceed with a discussion of some statistical methods for analysing time-series of spatial data.

2.4 TIME-SERIES ANALYSIS APPLIED TO SPATIAL DATA

If we have at our disposal a set of N x R observation $x(r, t)$ on variable $X(r, t)$, $r \in S; t \in T$, i.e. time-series for a variable as defined for different points in space, we may use some of the well-developed techniques of time-series analysis. A rich literature is available on this subject. We mention only a few valuable references: Brillinger (1975), Fishman (1969), Granger and Hatanaka (1964), Jenkins and Watts (1968), Kendall (1973), and Malinvaud (1966). For the use of spectral analysis with spatial data Granger (1969) contains a good exposition. In this section we consider briefly some time-series techniques that have proved to be operational in the context of spatial data.

The N x R observations will be regarded as realisations of the R-vector valued stochastic process (with real components) $X(t)$, defined by

$$X(t) = [X(1, t), \ldots, X(R, t)] \qquad , t \in T \tag{4.1}$$

For a further statistical analysis of properties of this process it is convenient to assume that the lower-order moments of this process provide us with an adequate description of its characteristics. We impose some a priori restrictions on the low-order moments, such that the process is covariance-stationary. For all $t \in T$ and r, $r' \in \{1, \ldots, R\}$ this implies

– the mean function does not depend on t, i.e.

$$\mathrm{E}[X(r, t)] = \mu(r, t) = \mu(r) \tag{4.2}$$

– the variance funtion does not depend on t, i.e.

$$\mathrm{var}\ [X(r, t)] = \gamma(r) \tag{4.3}$$

– the auto/cross covariance function depends only on the lag between points in time, i.e.

$$\mathrm{cov}\ [X(r, t)\, X(r', t')] = \gamma(r, r', t, t') = \gamma(r, r', \tau) \tag{4.4}$$

where $\tau = | t - t' |$

Expression (4.3) incorporates as a special case the auto-covariance function, i.e. when $r' = r$. This may be denoted by $\gamma(r, \tau)$. The cross-covariance and variance function may be used to define the auto/cross correlation function, i.e.

$$\rho(r, r', \tau) = \gamma(r, r', \tau) \, . \, \{\gamma(r) \gamma(r')\}^{-\frac{1}{2}} \tag{4.5}$$

The auto-correlation function ($r = r'$) will be denoted by $\rho(r, \tau)$.

Since many data-sets will not satisfy these assumptions, it is required to investigate properties of the data before starting the application of certain time-series techniques. In some cases a transformation of the original data may provide new data that will more likely satisfy the assumptions. Transformation to obtain a constant mean is a common practice, but transformation to obtain a homoscedastic series (i.e. constant variance in time) is much less common (and would indeed yield results which will be difficult to interpret). For further empirical applications one has to work with estimates of the functions defined above, which may be obtained by the method of moments. This yields estimates $\hat{\mu}(r)$, $\hat{\gamma}(r)$, $\hat{\gamma}(r, \tau)$, $\hat{\gamma}(r, r', \tau)$, $\hat{\rho}(r, \tau)$ and $\hat{\rho}(r, r', \tau)$ for different time-lags τ. Each of these estimates provides us with useful information on the time-series. The average gives a summary indicator of the level of each series, the variance indicates the instability of the series (sometimes the coefficient of variation may be preferred for this purpose). The cross-correlation for τ=0 gives a measure of closeness between the series for region r and that for region r'. In fact it amounts to the simple coefficient of correlation, defined in Section 2.3. Some insight in the temporal behaviour of a series is given by the correlogram, i.e. the picture of $\hat{\rho}(r, \tau)$ against τ. Comparison of these correlograms for the different regions enables us to detect common patterns in the data. A more comprehensive indication of such common characteristics and their leads or lags is provided by the cross-correlograms, i.e. the picture of $\hat{\rho}(r, r', \tau)$ against τ for each pair r, r'. These statistics have been used to study characteristics of regional unemployment series in Bartels (1977a, 1977b), and King et al. (1969).

Up till now we discussed useful statistics for studying time-series in the time-domain. However, it is often preferred to study the characteristics of the series in the frequency-domain, in order to obtain better insight into the periodicities that might be present in the series. One makes use of the property that under rather general conditions a stochastic process can be regarded as an infinite sum of periodical functions with different stochastic amplitude and frequency. In this case the variance of the time-series can be decomposed into independent additive components, which each express the contribution of the periodical functions defined for a certain frequency-interval to the variance of the series. This decomposition is commonly referred to as the spectrum of the series. For traditional spectral analysis covariance-stationarity of the series is necessary to make the standard tools applicable (new developments in spectral analysis allow for the study of non-stationary series). A standard procedure to estimate the spectrum starts from auto-covariance estimates, transforms these with particular weights to reduce the variance of the estimator, and applies a Fourier transform which then yields the spectrum estimate (cf. Jenkins and Watts (1968)). In this process several rather arbitrary decisions enter, viz. the choice of a method of trend-elimination if a time-trend causes the mean to be variable in time (Harkness, 1968, demonstrates that different trend-elimination procedures yield rather different conclusions), the number of auto-covariance estimates to be used, and the selection of the preferred weights. This shows that personal judgements are rather important in spectral analysis, which might quite well be considered as a severe drawback of this method. Another problem for application is that long time-series are required to make a useful application possible. In spatial analysis we have such series available for certain variables, and hence spectral analysis can be used to study such series. Several studies have been concerned with the structure of monthly or quarterly observed regional unemployment rates, e.g. Bartels (1977a, 1977b), Bartels et al. (1977), Bassett and Tinline (1969), Bassett and Haggett (1971), Hepple (1975), Cliff et al. (1975). Periodicities in spatially observed rainfall data are investigated in Tyson (1971) and daily changes in the quality of water in Thomann (1974). Bassett and Tinline (1969) also consider the spread of epidemics. In Bartels et al. (1977) we used spectral analysis to judge the importance of different periodical components (trend, business-cycle, season) in regional unemployment series and compared the results with those obtained with different techniques for an analysis in the time-domain. The conclusions for both approaches were rather similar which suggests that spectral analysis seems not to offer special advantages in an empirical context. Other time-series techniques allow for more flexibility with regard to alternative assumptions.

An extension of the analysis is given by cross-spectral analysis, which is concerned with the study of associations between pairs of correlated time-series. (A further extension is multivariate spectral analysis, which will not be discussed here.) It enables us to discover leads or lags between the periodical components as present in the series and to measure the degree of association between such components. The estimation technique is basically the same as above, but now cross-covariance estimates are the point of departure to calculate the cross-spectrum. Interpretable information is given by the squared coherency spectrum, which contains an indication of the degree of linear association between corresponding periodical components in both series. The phase spectrum may be used to discover leads or lags between corresponding components. Other informative statistics can be obtained as well. For spatial data analysis the associations between periodical components may yield useful information, e.g.

- they may suggest clusters of regions with similar characteristics. For example, in Bartels (1977a) we used cross-spectral analysis and were able to discover certain clusters of provinces with a common temporal behaviour.
- they may indicate a certain region as a leading one in a system of regions.

However, the same drawbacks as mentioned for univariate spectral analysis may apply and other, more flexible time-series methods (regression analysis with spatially and temporally lagged variables for example) may be more attractive.

Cross-spectral analysis compares all pairs of time-series in an isolated way. In Section 2.3 we noted that simple correlation analysis does the same, but that techniques are available to obtain a more informative presentation of all associations, ignoring possible time-lags. It is possible to extend the factor-analytical approach discussed before to the present analysis in the frequency domain, where we take possible time-lags explicitly into account. The procedure amounts to using the output of spectral analysis in a principal components analysis in order to reveal common components present in all the series (R) being investigated. In Brillinger (1975) an extensive discussion of this approach is given. It implies a rather rigorous extension of the concept of unobservable variables. The crucial assumption is that each element of a stochastic process can be explained as a linear combination of unobservable variables (common factors) both lagged and non-lagged. One looks for the

P-vector valued unobservable stochastic process $Y(t)$, with as elements the common factors, which describes the R-vector valued observable stochastic process $X(t)$ in some optimal way. The stochastic processes are assumed to be linked by means of linear, time-invariant operations, i.e.

$$X(t) = \alpha + \sum_{\tau} B(\tau)\, Y(t\text{-}\tau) + \epsilon(t) \qquad , t \in T \tag{4.6}$$

with

- α an R-vector of coefficients
- $B(.)$ an R x P matrix of coefficients. For each time-lag τ such a matrix exists
- $Y(.)$ the P-vector of common factors
- $\epsilon(.)$ an R-vector of disturbances.

Compared with the factor analytical model (3.3) the extension amounts to incorporating lagged values of the common factors in the analysis, so that each observation $x(r, t)$ may be regarded as a linear combination of contemporaneous and lagged common factors and a disturbance term. Like in conventional factor analysis one has to choose some optimality criterion. This may consist in determining the real-valued series $y(i, t)$, $i \in \boldsymbol{P} = \{1, \ldots, P\}$ that has a maximum variance and zero coherency with the real-valued series $y(j, t)$, $j \in \boldsymbol{P}$, $j < i$, for $j = 1, \ldots, P$. Brillinger (1975) has shown that the results of spectral analysis can be used to generate inter alia the spectra of the unobservable variables, and to discover phase-shifts between unobservable and observable variables. For the analysis of time-series recorded at different points in space, this latter approach makes it possible to detect common patterns underlying the structure of all series, while taking lagged associations into account. However, the technique is rather new and little experience in an empirical context is available. An application with spatial data is given in Bartels (1977a). There we obtained results that did not extend the insight derived from the traditional spectral-analysis much, while we felt also the need of more statistical testing procedures which are lacking in this context.

Above we discussed some time-series techniques and their applicability in view of getting a better description of the regularities of spatial data. The results obtained by means of these techniques may also serve to provide suggestions for the specification of simple models, which are able to generate the observed regularities, and may consequently be used for prediction and policy-evaluation purposes. We discus two important types of such models, i.e. (cf. Bennett and Haining, 1976):

– general *transfer function models*, specified by (deleting the spatial index):

$$X(t)+\alpha_1 X(t\text{-}1)+....+\alpha_a X(t\text{-}a) = \beta_0\, Y(t)+....+\beta_b\, Y(t\text{-}b) \qquad (4.7)$$

where X denotes a dependent variable
Y is an independent observable variable
a and b indicate the maximum number of lags.

This model is determined by a transfer function of order a, b.
– an *autoregressive moving-average model*, specified by

$$X(t)+\lambda_1 X(t\text{-}1)+ .... + \lambda_c X(t\text{-}c) = \delta_0\, \epsilon(t) + ... + \delta_d\, \epsilon(t\text{-}d) \qquad (4.8)$$

where ϵ denotes a stochastic unobservable variable
c and d indicate the maximum number of lags.
This model may be referred to as an ARMA (c, d)-model.

For models with relatively small lags important properties of their theoretical characteristics are known. These can be compared with several empirical statistics for a given time-series. Two points of departure could be chosen for this type of analysis:

– analysis of properties in the *time-domain*.
This requires study of the estimated autocorrelation, partial autocorrelations and cross-correlations. Their properties are compared with those of theoretical models in order to suggest an appropriate specification for the phenomena under study (This procedure is also known as the Box-Jenkins technique).
– analysis of properties in the *frequency-domain*.
In this case properties of the spectra, gain and phase form the point of departure to derive suggestions for a simple model specification.

A third approach involves simple least-squares estimation of several simple models and selecting a best model by means of some explicit criteria. A full discussion of these approaches is contained in Bennett (1974), where also several empirical examples are given. From this discussion it appears that the third approach seems quite attractive as compared with the other alternatives, for its simplicity and for the absence of important subjective judgement. However, Bennett suggests to use them in combination to obtain an unambiguous model specification.

It is obvious that this type of analysis succeeds only in giving a empirical description of the phenomenon concerned, without giving attention to an explanatory theory underlying this empirical behaviour. Since it gives explicit rules which govern the values of the variable, the procedure is helpful for prediction purposes. This is only the case, however, if no important structural changes occur. Modelling the impact of such changes requires a more explanatory approach to model-building, as to be discussed later in this contribution.

Having discussed the use of various time-series techniques which have proven to be of some operational value for the analysis of spatial data, we shall now investigate how these techniques have been adapted to take the spatial nature of the data better into account.

2.5 ADAPTATIONS OF TIME-SERIES ANALYSIS TO THE SPATIAL CONTEXT

In this section we consider some approaches that have attempted to link up more explicitly with the spatial nature of the data. Let us start with the simple case of a cross-section of spatial data $x(r)$, $r=1, \ldots, R$ at a certain instant in time. As with stationary time-series one may want to know to what extent autocorrelation can be considered as statistically significant. For time-series analysis we indicated estimates of the autocorrelation as measures of temporal autocorrelation. But it appears that, if one additionally assumes the stochastic process to be normally distributed, the distribution under the null hypothesis of no autocorrelation is rather complex (cf. Malinvaud, 1966). Therefore, other test-statistics have been proposed. Well-known examples are the Von Neumann ratio and the Durbin-Watson statistic.

For spatial series one has looked for some analogies. Therefore we need some assumptions analogous to those for time-series, viz.

$$\mathrm{E}(X(r)) = \mu \tag{5.1}$$

$$\mathrm{var}\,(X(r)) = \sigma^2 \tag{5.2}$$

$$\mathrm{E}(X(r)\,X(r'_{kr})) = \gamma(k) \tag{5.3}$$

In (5.3) k denotes a spatial lag and r'_{kr} regions that are k-order contiguous to region r (cf. Hordijk, 1974, for a definition of this term). First it should be noted that these assumptions are again quite restrictive ones. A constant mean as required by (5.1) is in general not realistic. For example, if one studies regional unemployment one may expect the mean to be higher in regions with structural employment problems than in those with stable near-full employment. Variations in the variance may be expected as well. Finally, a diversified system of transport routes will result in differences in the degree of interactions among regions, as expressed in (5.3). Some adaptation of the data may therefore be needed, e.g. one could work with residuals of a regression which rules out systematic influences on the level of a variable. Another possibility is to use some transformation of the data, to obtain constant mean and/or variance. Condition (5.3) is more difficult to satisfy, however.

Several measures of spatial autocorrelation have been proposed, all implicitly using the conditions as stated above. Geary (1954) proposed a measure of first order spatial autocorrelation which is an extension of the Von Neumann ratio. Moran (1950) used the temporal autocorrelation coefficient and adjusted this for spatial data. Following Dacey (1968), Cliff and Ord (1969), (1973) have proposed a generalisation of Moran's coefficient which does not use assumption (5.3). It implies the use of non-binary weights applied to observations of contiguous regions, in order to introduce varying levels of spatial interaction in the measurement. Assumption (5.3) is then replaced by

$$\mathrm{E}(X(r)X(r'_{kr})) = \gamma(k, w(r, r'_{kr})) \tag{5.4}$$

with $w(r, r'_{kr})$ a non-binary weight. If we express these a priori fixed weights in a matrix $\boldsymbol{W}$ and the observation in an R-vector $\boldsymbol{x}$ the generalised Moran coefficient, apart from a constant, equals

$$\mathrm{GMC} = \frac{(\boldsymbol{x}\text{-}\boldsymbol{x})' \boldsymbol{W} (\boldsymbol{x}\text{-}\boldsymbol{x})}{(\boldsymbol{x}\text{-}\boldsymbol{x})'(\boldsymbol{x}\text{-}\boldsymbol{x})} \tag{5.5}$$

with $\boldsymbol{x}$ a vector with as elements the mean of the observations and $\boldsymbol{W}$ having non-zero elements for continguous regions only (different orders of contiguity are allowed in this definition). Cliff and Ord also succeed in deriving asymptotic moments, which can be used for hypothesis testing. Furthermore, the asymptotical distribution of GMC is a normal one under the null-hypothesis of independence given some additional assumptions. However, Sen (1976) claims that their proof of asymptotic normality is erroneous, and presents an alternative proof of this property. The use of a matrix of weights in this kind of descriptive statistical analysis has also evoked some discussion. Several options have been explored for the specification of the elements of matrix $\boldsymbol{W}$, like indicators of the relative surface of a region, the relative common boundary, the distance between centers of regions and combinations of these indicators (cf. Dacey, 1968; and Chatterji, 1976). An a priori best choice seems impossible. In a recent contribution Kooijman (1976) proposes to choose weights so as to maximise Moran's coefficient with certain constraints on the weights. The appeal of this procedure is not very clear, however. A drawback of the application of non-binary weights to observations of contiguous regions consists of the fact that this introduces notions related to the explanation of spatial dependencies in the exploratory analysis. This would be better justified if appropriate tests were included to judge whether such dependencies are appropriately represented by the specific choice of matrix $\boldsymbol{W}$. As long as such tests are not available the use of binary weights seems to be quite well defendable. If one still wants to correct for influences like the size of areal units on the observations, the use of weighted *observations* could be preferred (cf. also Chatterji, 1976). With regard to Geary's coefficient we have to mention Lebart (1969) who generalises this coefficient for higher order contiguity and derives the appropriate moments. By means of some numerical examples it is suggested that this coefficient will be asymptotically normally distributed in empirical applications.

As a special use of spatial autocorrelation coefficients we consider tests for such autocorrelation among regression disturbances in a cross-regional linear regression model. The effects of ignoring autocorrelation among disturbances are well-known; for a discussion related to spatial applications we may refer to Martin (1974). Several recent contributions have proposed tests on such autocorrelation, for example, Cliff and Ord (1970), (1972), (1973), Fisher (1971) and Hordijk (1974). The tests are mostly restricted to the situation where the null-hypothesis of no-spatial autocorrelation is tested against the alternative hypothesis of a generalised first-order spatial Markov scheme describing interactions among disturbances, i.e.

$$H_1 : \boldsymbol{\epsilon} = \rho \boldsymbol{W} \boldsymbol{\epsilon} + \boldsymbol{\xi} \tag{5.6}$$

where

$\boldsymbol{\epsilon}$ is the R-vector of disturbances in the cross-regional model
$\rho \in [0,1]$ is a spatial autocorrelation parameter
$\boldsymbol{\xi}$ is an R-vector of mutually independent, normally distributed variables with zero mean and constant variance.

(A more general alternative hypothesis, allowing for higher-order spatial connections may be found in Hordijk, 1974.)

In order to apply a test-statistic like (5.5) to judge autocorrelation in the disturbances in an L.S. regression framework, we need estimates of these unobservable disturbances. Some alternatives can be mentiond, viz:

1. Use OLS residuals. These have a non-scalar covariance matrix under the null-hypothesis, which makes the moments of GMC dependent on the regressor matrix. If the moments are calculated, an asymptotically $N(0,1)$ distributed test-statistic is easily obtained.
2. Use linear unbiased (LU) estimators that do possess a scalar covariance matrix. These are the so-called LUS residuals, which have the advantage of yielding moments that do not depend on the regressor matrix. Possible members of this class of LUS residuals are the BLUS residuals (best LUS residuals; best in the sense that they satisfy a certain optimality criterion. Cf. Koerts and Abrahamse, 1971; Theil, 1965, 1971) and the recursive LUS residuals as proposed recently by Philips and Harvey (1974) and Brown et al. (1975).

3. Use linear unbiased estimates that possess a fixed covariance matrix, which may be regarded as typical for a certain class of data. Hence, the calculation of the moments needs not to be repeated each time if data belonging to this class are analysed. In temporal models the periodicities present in most time-series allow for the selection of a meaningful representative matrix, cf. Abrahamse and Koerts (1971) and Dubbelman et al. (1972). In the present context of spatial models this seems more difficult and the procedure will be of little operational use.

A different approach in L.S. analysis consists of estimation of autocorrelation parameter ρ. This may be done along several different lines. Substituting the Markov scheme in the hypothesized relation yields an estimable autoregressive scheme which is, however, non-linear in the parameters. A second method is a least-squares regression on the residuals $\boldsymbol{e}$ of the following type:

$$\boldsymbol{e} = \rho \boldsymbol{W} \boldsymbol{e} + \boldsymbol{\xi} \tag{5.7}$$

The estimate $\hat{\rho}$ is used to transform the original data; afterwarth a new regression is performed which yields new residuals. An iterative procedure may be followed to derive the final value of $\hat{\rho}$ (cf. Bodson and Peeters, 1975 for a discussion of drawbacks of this procedure. They propose a scanning procedure which tries different values of $\hat{\rho}$ and chooses that one which minimises the sum of squared residuals).

Maximum likelihood estimation provides still another approach to the present problem. It enables us to obtain simultaneously estimates of the regression parameters and the autocorrelation parameter (cf. Hepple, 1976; Ord 1975; Hordijk and Paelinck, 1976). Autocorrelation may be tested by using the asymptotical normal distribution property of the estimated autocorrelation parameter or by using a likelihood ratio test (compare e.g. Goldfeld and Quandt, 1972 for such tests).

Since all test statistics use asymptotical properties it is important to study their small sample behaviour. This has been done by means of Monte Carlo simulation. For the Moran coefficient using binary weights Bartels and Hordijk (1977) investigate the properties for the different residuals. OLS residuals were found to yield the highest power. Brandsma and Ketellapper (this volume) have extended their analysis by incorporating also the M.L. approach and studying additional characteristics of the test-statistics. This study did not change the previous conclusion.

If application of such a test-statistic suggests presence of significant autocorrelation, a better specification of the model is required. One may use a generalised least-squares approach which implies transformation of the variables (cf. inter alia Bodson and Peeters, 1975, Fisher, 1971, Hordijk, 1974 and McCamley, 1973). In Martin (1974) it is demonstrated by means of a simulation experiment that the use of simple spatial first differences (i.e. observation–weighted average of observations in contiguous regions) is already an improvement to OLS applied to the original data. Other possible remedies are the incorporation of different independent variables and the introduction of spatial lags in the regression.

A straightforward generalisation of the contiguity coefficients considered up till now leads to the definition of similar coefficients for pairs of variables analogous to the cross correlation coefficient in temporal analysis. Lebart (1969) constructed a matrix of bivariate contiguity coefficients for different orders of contiguity, using a generalisation of the univariate Geary approach. The elements of this matrix express the strength of association between a pair of variables observed in regions which are contiguous of a certain order. Such a matrix may again be compactly summarised using factor analysis, although the results will not be easy to interpret. Of course, a similar generalisation of Moran's coefficient is possible. But at present such bivariate contiguity coefficients seem little informative, since little is known about the range of their values and the first-order moments.

A generalisation that goes still further combines bivariate spatial correlation analysis with temporal autocorrelation analysis. This starts from observations of the type $x(r, t)$ and $y(r, t)$, $r=1, \ldots, R$, $t=1, \ldots, T$. Analogous to the previous discussion one might construct bivariate spatial-temporal correlation coefficients that measure the strength of association between a pair of variables, as observed in contiguous regions with a certain time-lag. Such generalisations are proposed in Hordijk and Nijkamp (1976) although the operational value of such coefficients is limited, since little is known about their statistical properties.

The discussion above has been concerned with a cross-section of spatial data, while we have taken the spatial geometry only partly into account. A more complete treatment of the element of space requires the use of spatial co-ordinates.

A first case consists in the analysis of a variable $X(r)$, as defined for a line in space with co-ordinates r defined for regular intervals. Assuming constant mean and variance, and cross-covariance depending only on the spatial lag, one might use analogies of the univariate time-series analysis. If one expects spatial periodicities, one-dimensional spectral analysis could be used to investigate this further. Tobler (1969) applied this technique to investigate the population density (which had to be approximated using hypothesized models) along U.S. Highway 40 in order to detect regularities in settlement patterns. A criticism on this study is given in Rayner and Colledge (1973) who believe that the spectrum might be generated by a random process; the use of normalised input data is proposed as a better alternative.

A second case accounts explicitly for two spatial dimensions. One studies variables of the type $X(r_1, r_2)$ with r_1 and r_2 uniformly spaced co-ordinates. If observations on such variables are not directly available it may be possible to construct them given a number of point observations in space. Application of time-series ideas is facilitated by the assumption that $X(r_1, r_2)$ is a two-dimensional covariance-stationary spatial stochastic process. The assumption amounts to

$$E[X(r_1,r_2)] = \mu \tag{5.8}$$

$$\text{var}\,[X(r_1,r_2)] = \sigma^2 \tag{5.9}$$

$$E[(X(r_1,r_2) - \mu)(X(r_1+k, r_2+m) - \mu)] = \gamma(k,m) \tag{5.10}$$

In (5.10) k and m denote spatial lags. $\gamma(k, m)$ is the autocovariance function, which may be used to define the autocorrelation function $\rho(m, k) = \gamma(m, k)/\sigma^2$. Estimates of this latter provide elements of a spatial autocorrelation matrix. Again Fourier transforms can be used to decompose the variance of such two-dimensional series into components associated with regular spatial wave-like functions, into two directions. This type of analysis is the subject of two-dimensional spectral analysis, cf. Bartlett (1964), (1975), (1976), Barton and Tobler (1971), and Rayner and Colledge (1972).

If a pair of variables is studied while both may be assumed to represent a two-dimensional covariance stationary process then the cross-covariance function may be the point of departure. It equals

$$E[X_1(r_1, r_2) - \mu_1)(X_2(r_1 + k, r_2 + m) - \mu_2)] = \gamma_{12}(k, m) \quad (5.11)$$

with μ_1 and μ_2 the respective mean functions. Bivariate two-dimensional spectral analysis seems to have received little attention, however, and operational tools are not available in this context.

A third case accounts explicitly for two spatial dimensions and a temporal dimension. So one studies variables of the type $X(r_1, r_2, t)$. The assumptions amount to

$$E[X(r_1, r_2, t)] = \mu \quad (5.12)$$

$$\text{var}(X(r_1, r_2, t) = \sigma^2 \quad (5.13)$$

$$E[(X(r_1, r_2, t) - \mu)(X(r_1 + k, r_2 + k, t + \tau) - \mu) = \gamma(k, \tau) \quad (5.14)$$

(5.14) represents the spatial-temporal autocovariance function, which can be used to define the spatio-temporal autocorrelation function. For a pair of variables, the spatio-temporal cross-covariance function may be defined by

$$E[(X_1(r_1, r_2, t) - \mu_1)(X_2(r_1 + k, r_2 + k, t + \tau) - \mu_2)] = \gamma_{12}(k, \tau) \quad (5.15)$$

which may again be used to define the spatio-temporal cross-correlation function. Estimates of these spatio-temporal correlations can be obtained in an analogous way as in the purely temporal case. The results could again serve to suggest a simple spatio-temporal autoregressive-moving average model, as has been demonstrated in Bennett (1975a).

It is also possible to define spectra for the spatio-temporal processes defined above. This topic is treated in Bartlett (1975), (1976). However, much of the discussion seems mainly of theoretical value and not easily applicable to the data we have commonly at our disposal in spatial analysis.

With regard to two-dimensional spatial analysis it is frequently remarked that the required assumptions will be less easily met than in time-series analysis (cf. inter alia Berry, 1971, for this view). However, also in a spatial context transformation of data is tractable in order to obtain series that might better satisfy the assumptions. Estimation of trend-surfaces by means of low order polynomials in the spatial co-ordinates can be mentioned as a tool to eliminate spatial trends (cf. Berry, 1971; Cliff et al., 1975). Alternatively, first spatial differences could be used to obtain a stationary series (cf. Bennett, 1975; Huijbregts, 1975). A more serious drawback for operational applications is the difficulty to associate a meaningful interpretation with spatial periodical functions, since we will not easily expect such functions with approximately constant amplitude and frequency to exist in real space. There have been attempts towards an interpretation, using a subdivision into spatial scale elements, viz. national, regional and local components (cf. Tobler, 1973). However, results obtained up till now along these lines do not yield very informative results in an empirical context. For a more elaborate treatment of statistical problems with two-dimensional analysis, like auto-regressive processes, we refer to Besag (1974).

Having discussed several methods which can be considered as tools in an exploratory analysis of spatial data, we now turn to a discussion of a more explanatory approach to the analysis of spatial data.

2.6 SINGLE EQUATION EXPLANATORY MODELS

Traditional econometric analysis is sometimes thought to proceed along the following stages: theory formulation, data gathering, estimation of the parameters in the model and testing their significance, ex-post validation of the complete model, ex-ante validation (i.e. beyond the sample), and simulation with the model (for forecasts and policy-evaluation). In empirical econometrics the procedure may be less straightforward. As noted before, a clear-cut a priori theory is frequently absent while also appropriate data for hypothesis-testing may not be available. In such situations the order of the stages may differ, and exploratory statistical analysis as discussed above may be of use in the stage of theory formulation. For the discussion in this and the following section we shall assume that the first two stages have been performed. So we start from the situation where a full explicit causal model formulation is available, and shall consider the subsequent stages of the statistical analysis. In this section the single-equation model will be treated while we focus in the next section on simultaneous models.

2.6.1 Cross-sectional analysis with spatial data

Single equation models have been frequently developed for testing with a cross-section of spatial data. As an example we mention attempts to associate characteristics of intra-regional income distributions with several intra-regional socio-economic variables. This kind of cross-sectional regression analysis has to take the spatial nature explicitly into account. This may require testing for spatial autocorrelation of disturbances along the lines set out above. Furthermore, the possibility of spatial-heteroscedasticity of disturbances (i.e. a non-constant variance) has to be considered. In the time-series context several tests for the latter are available. A good summary of constructive and non-constructive tests is contained in Goldfeld and Quandt (1972, Chapter 3). A constructive test implies that if it rejects the null hypothesis of homoscedasticity it provides also a direct method of estimation which takes account of the heteroscedasticity. A non-constructive test only establishes the presence or absence of heteroscedasticity and does not suggest and estimation procedure. As non-constructive tests, F-tests based on BLUS residuals (Theil, 1971) or recursive LUS residuals (Harvey and Philips, 1974) may be mentioned. Constructive tests are mainly obtained by using a likelihood ratio test procedure, cf. Harvey (1976) and Merkies (1975) for example. Although testing for heteroscedasticity is not that common in cross-regional analysis, the approaches used in the temporal context can be easily applied to spatial data. If heteroscedasticity is found to exist, a weighted least-squares approach will have to be followed (cf., for example, Bartels, 1977a, for an explanation of regional incomes and Kau and Lee, 1977 for estimation of the urban population density gradient).

As a special member of this class of models we mention spatial autoregressive models. A mixed regressive first-order spatial autoregressive model may be written as

$$y = X\alpha + \rho Wy + \epsilon \tag{6.1}$$

with

– $\boldsymbol{y}$ an R-vector of observations on the dependent variable
– $\boldsymbol{X}$ an R x I matrix of observations on independent variables
– α an I-vector of parameters
– ρ the spatial autocorrelation parameter introduced before
– $\boldsymbol{W}$ an R x R matrix of weights
– $\boldsymbol{\epsilon}$ an R-vector of disturbances

Since OLS estimates of the parameters can be shown to be inconsistent, alternative estimation procedures like maximum likelihood have to be followed (cf. Ord, 1975). An extension of model (6.1) is given by a mixed regressive-autoregressive-moving-average model of the form

$$y = X\alpha + \rho_1 Wy + \epsilon + \rho_2 A\epsilon \tag{6.2}$$

with A an R x R matrix of weights.

A discussion of estimation procedures for such a model can be found in Cliff and Ord (1975).

2.6.2 Combination of time-series and cross-section analysis
In some cases one may be lucky to posses a time-series of spatial data, i.e. observations $x(r, t)$ for $r = 1,..., R$ and $t = 1,..., T$. Observations on such variables may be used simultaneously in an explanatory analysis, where different assumptions can be made. We mention briefly two possiblities (cf. also Johnson and Lyon, 1973; Paelinck and Wagenaar, 1976).

1. A *fixed effects* model, which assumes a fixed influence of time and space on some of the coefficients. For example, if this is applied to the constant term only, we obtain the following model:

$$y(r, t) = \alpha + \beta(r) + \gamma(t) + x'(r, t)\delta + \epsilon(r, t) \tag{6.3}$$

   with $x(r,t)$ a vector of independent variables and $\epsilon(r,t)$ satisfying traditional assumptions (independency, homoscedasticity, etc.). This model can be estimated using all observations in a pooled regression analysis.
2. A *random effects* model, which assumes constant coefficients, i.e.

$$y(r, t) = x'(r, t)\delta + \epsilon(r, t) \tag{6.4}$$

   but introduces the specific effects of time and space as random variables in the disturbances. Hence, this model is also called an error-component model. It has been used by Kuh (1959) and Nerlove (1971). In the latter it is assumed that the disturbances can be decomposed into uncorrelated components, i.e.

$$\epsilon(r, t) = u(r) + v(t) + w(r, t) \tag{6.5}$$

with

$$\begin{aligned}
E[u(r)\,u(r')] &= \sigma_1^2 && r = r' \\
&= 0 && r \neq r' \\[1em]
E[v(t)\,v(t')] &= \sigma_2^2 && t = t' \\
&= 0 && t \neq t' \\[1em]
E[w(r,t)\,w(r',t')] &= \sigma_3^2 && r = r', \quad t = t' \\
&= 0 && r \neq r', \quad t \neq t'
\end{aligned} \tag{6.6}$$

These assumptions suffice to develop an operational estimation procedure, like generalised least-squares or maximum-likelihood estimation. Application of this procedure requires the error components to be uncorrelated with the independent variables. If this assumption is thought to be not valid, the first method can best be used.

The choice between both type of models depends on whether one is interested in estimating separately the influence of time and space on the dependent variable or not. If one wants to estimate only the parameter-vector δ, the second approach could be followed. But if one wants also an estimate of the specific temporal and spatial components, the first method should be followed.

The two examples given above represent only part of the possible use that can be made of a cross-section of time-series data in an explanatory analysis. Alternative specifications could introduce fixed or random temporal and spatial components in the elements of vector δ. Since more time-series of spatial data are becoming available it seems worthwhile to pay more attention to this type of analysis than has been done up till now in regional science (an application has been given in Glejser et al., 1973, for the estimation of a regional econometric model).

2.6.3 *Traditional spatial interaction models*

A special class of single equation models deserves some extra attention for its frequent use in spatial analysis and some typical statistical problems associated with it. This class contains models to explain inter-regional flows of goods, persons, traffic, etc., and which are called spatial interaction models (sometimes terms like gravity or potential models are used as well). An introduction to such models is contained in Wilson (1974), while the calibration of such models is considered in inter alia Batty et al. (1973). For such models the dependent variable may be written as $Y(r, r')$, i.e. the flow from area r to area r'.

A traditional specification of an interaction model assumes that it makes sense to model the associationship analogous to Newton's gravity law. A generalised deterministic representation of such interaction models is

$$Y(r,r') = \psi_1\, \alpha_1(r,r')) \,.\, \psi_2(\boldsymbol{x}(r), \alpha_2) \,.\, \psi_3(\boldsymbol{z}(r'), \alpha_3) \,.\, \psi_4(d(r,r'), \alpha_4) \tag{6.7}$$

with

$\psi_1(\ )$	a function of parameters which may be required to make the model consistent (cf. below)
$\alpha_1(\ )$	a parameter vector
$\psi_2(\boldsymbol{x}(r), \alpha_2)$	a multiplicative function of push variables $\boldsymbol{x}(r)$ with associated parameters α_2
$\psi_3(\boldsymbol{z}(r'), \alpha_3)$	a multiplicative function of pull variables $\boldsymbol{z}(r')$ with associated parameters α_3
$\psi_4(d(r,r'), \alpha_4)$	a function of a measure of the cost to move from r to r'. Frequently distance is used to approximate this variable. (The use of a standardised distance measure is sometimes recommended, e.g. Johnston, 1976 proposes to use the ratio of distance to the range of distances in order to minimise the influence of the geometry of the spatial system. In Masser, 1977 distance divided by mean trip length is used.)

With respect to function $\psi_4(\ )$ several a priori choices can be made, such as a negative power function, an exponential function, a Tanner function (which combines the former two), a modified Weibull function, incomplete gamma or beta functions, and a combination of functions into a family of deterrence functions so that parameters may differ for sub-sets of interactions (cf. Openshaw (this volume) for a fuller discussion). In application of models like (6.7) an important distinction can be made between constrained and unconstrained models. We consider first the unconstrained models briefly.

The unconstrained model makes use of one or more independent variables at the right-hand side of (6.7). $\psi_i(\alpha_1(r, r'))$ simply reduces to a single parameter α_1 and the model with exponential deterrence function becomes

$$Y(r,r') = \alpha_1 \, x_1(r)^{\alpha_{21}} \ldots x_I(r)^{\alpha_{2I}} z_1(r')^{\alpha_{31}} \ldots z_J(r')^{\alpha_{3J}} \exp\{-\alpha_4 d(r,r')\} \tag{6.8}$$

For purposes of statistical estimation a stochastic disturbance has to be introduced. With respect to the unconstrained model (6.8) we mention two possibilities, viz.

1. an additive error term, i.e.

$$y(r,r') = Y(r,r') + \epsilon(r,r') \tag{6.9}$$

with $\epsilon(r, r') \sim N(0, \sigma^2)$ and mutually independent

2. a multiplicative error term, i.e.

$$y(r,r') = Y(r,r') \cdot \epsilon(r,r') \tag{6.10}$$

with $\epsilon(r, r')$ mutually independent and having a log-normal distribution with mean $\exp\{\frac{1}{2}\sigma^2\}$ and variance $\exp\{2\sigma^2\} - \exp\{\sigma^2\}$. Alternatively one might use instead of $\epsilon(r, r')$ the term $\exp\{\nu(r,r')\}$ with $\nu(r, r') \sim N(0, \sigma^2)$, cf. for example Cesario (1975).

Let us first consider specification (6.10). The specific assumptions are motivated by the desire to be able to transform a non-linear model into a linear one, so that its parameters can be estimated by means of traditional linear regression analysis. For a model as (6.10) the logarithmic transformation yields a model linear in the parameters (except for α_1) and with normally distributed error term. Hence, classical regression analysis can be used to estimate the parameters $\alpha_{21}, \ldots, \alpha_{2I}, \alpha_{31}, \ldots, \alpha_{3J}, \alpha_4$, which estimates possess convenient properties (i.e. unbiased, normally distributed, etc.). In order to obtain an unbiased estimate of α_1 the unbiased estimate of ln α_1 can be used, taking into account an appropriate multiplication factor (cf. Heien, 1968). Another feature of model (6.10) has to be noted. Having obtained estimates from the log-linear model one will often be interested in obtaining an unbiased predictor of the dependent variable. If we denote the OLS parameter estimates in the log-linear model by ln $a_1, a_{21}, \ldots, a_{2I}, a_{31}, \ldots, a_{3J}, a_4$ the following predictor is frequently used:

$$\hat{Y}(r, r') = \exp \{ \ln a_1 + \sum_{i=1}^{I} a_{2i} \ln x_i(r) + \sum_{j=1}^{J} a_{3j} \ln z_j(r') - a_4 \, d(r, r') \} \qquad (6.11)$$

with $x_i(r)$ and $z_j(r')$ observations on exogenous variables. Expression (6.11) however is a biased predictor. This will be clear if we consider the conditional mean

$$\begin{aligned} \mathrm{E}(Y \mid \text{exogenous variables}) &= Y(r, r') \,.\, \mathrm{E}(\epsilon(r, r')) \\ &= Y(r, r') \,.\, \exp \{ \tfrac{1}{2} \sigma^2 \} \qquad (6.12) \end{aligned}$$

Expression (6.11) gives a prediction of $Y(r, r')$, while (6.12) indicates that some multiplication factor will have to be applied to (6.11) in order to get an unbiased predictor of Y. The appropriate factor is rather complicated and may be found in Bradu and Mundlak (1970) and Goldberger (1968).

Expression (6.12) may cause some doubt on the appropriateness of the assumptions on the error term, since it could be preferred to have a conditional mean that equals the deterministic part of the relation, as is commonly the case with additive error terms. Alternative assumptions have been proposed to circumvent this drawback. For example, Teekens and Koerts (1972) assume $\epsilon(r, r')$ to be lognormally distributed with $\mathrm{E}(\epsilon(r, r')) = 1$ and var $(\epsilon(r, r')) = \sigma^2$. From (6.12) we see that in this situation the conditional mean equals $Y(r, r')$. A complication arises, however, since in the transformed model we have $E(\ln \epsilon(r, r')) = \frac{1}{2}\sigma^2$, so that OLS would result in biased parameter estimates. A solution exists in using maximum likelihood procedures (cf. Teekens and Koerts, 1972) or rewriting the model, i.e. incorporating $-\frac{1}{2}\sigma^2$ in the constant term (cf. Heien, 1968). These notions on estimation and forecasting with log-linear models are of extreme importance, also for the class of spatial interaction models considered here. Therefore, more attention has to be given to these kinds of complications in spatial applications. At present, such attention is virtually absent.

After these remarks with regard to specification (6.10) we now consider some implications of (6.9). It is clear that parameter-estimation requires some non-linear estimation procedure. Maximum likelihood is an attractive method. If the derivatives of the log-likelihood function cannot be analytically obtained a numerical maximisation procedure will have to be followed. The M.L. method provides an asymptotic expression for the standard errors of the estimates, using the information matrix. Estimation of this matrix requires the analytical calculation or numerical approximation of second order derivatives. For tests of alternative hypotheses the likelihood ratio provides a flexible tool (more details on maximum likelihood estimation may be found in Goldfeld and Quandt, 1972). Alternative non-linear methods are also available. These use mainly an approach that aims to minimise some measure of divergence between observed and predicted values of the dependent variable. As a measure of divergence some special case of the (weighted) Minkowski or L_p norm may be chosen. This latter equals (cf. Ronner, 1977, for the use of the L_p norm in a linear regression framework):

$$L_p = \{\sum_r \sum_{r'} |y(r, r') - Y(r, r')|^p\}^{\frac{1}{p}}, p \geqq 1 \qquad (6.13)$$

The estimation procedure consists of minimising the norm for varying parameter-values. Best-known is the use of the L_2 norm, which amounts to a least-squares approach. An advantage of this norm is that it enables us to calculate asymptotic standard errors, cf. for example Mizon (1977). In spatial interaction analysis the L_1 norm is sometimes used, and also the L_∞ or Tchebycheff norm defined by

$$L_\infty = \max_{r, r'} |y(r, r') - Y(r, r')| \qquad (6.14)$$

Sometimes a weighted least-squares approach is preferred, e.g. if the chi-square metric is used. It is difficult to indicate a priori the preferred minimisation criterion. Rice and White (1964) discuss some different L_p norms and argue that the choice depends on the distribution of the errors. Using some numerical examples they demonstrate that the L_1 norm is attractive for error distributions with long tails (i.e. outliers in the data, cf. also Ronner, 1977), L_∞ for distributions with sharply defined extremes (e.g. uniform distribution) and L_2 for distributions somewhere between these extremes (e.g. a normal distribution). A basic difficulty is of course that a priori little is known about the error distributions. Therefore, it seems more important to judge the alternative estimation procedures with regard to the possibility to obtain standard errors of estimates. M.L. and non-linear least-squares will then result as the preferred estimation methods. In many empirical studies one simply minimises some divergence measure, without attention to significance testing of parameters. For example, Openshaw (1976) considers several alternatives and indicates non-linear least-squares as yielding the best fit, but in Openshaw (this volume) it is concluded that the influence of estimation method on goodness-of-fit is not remarkable. Such studies focus on a single scalar measure of goodness-of-fit. Batty and March (1976) have shown that it may make more sense to investigate the original residuals in order to improve the fit.

We now turn to a discussion of the constrained interaction models. Such models are subdivided into two types, i.e. singly and doubly constrained models. For singly constrained models one requires

$$\sum_{r'} \hat{y}(r, r') = \sum_{r'} y(r, r') \equiv y(r, .) \tag{6.15}$$

and

$$\sum_{r'} Y(r, r') = y(r, .) \tag{6.16}$$

with

$\hat{y}(r, r')$ the predicted flow from r to r'

$y(r, r')$ the observed flow from r to r'

Alternatively, the summation may be taken over r and set equal to $y(.,r')$, i.e. the total number of flows arriving in r'. If both requirements are imposed, we have a so-called doubly constrained interaction model. We shall discuss the singly constrained model, as determined by (6.15) and (6.16) further. Let us start from the following simple interaction model:

$$Y(r, r') = \psi_1(\alpha_1(r, r'))\, x_1\,(r)^{\alpha_{21}} z_1(r')^{\alpha_{31}} \exp\{-\alpha_4\, d(r, r')\} \tag{6.17}$$

Using constraint (6.16) we may write model (6.17) as

$$Y(r, r') = y(r, .)\ \frac{x_1(r)^{\alpha_{21}}\ z_1(r')^{\alpha_{31}} \exp\{-\alpha_4\, d(r, r')\}}{\sum_{r'} x_1(r)^{\alpha_{21}} z_1(r')^{\alpha_{31}} \exp\{-\alpha_4\, d(r, r')\}} \tag{6.18}$$

The predicted flows are then obtained by estimating the parameters and substituting these into (6.18); these values are denoted by $\hat{y}(r, r')$.

For estimation purposes one has to introduce a disturbance term in (6.18). This may again be proceeded along several lines.

1. An additive error term like in (6.9). Use non-linear least-squares to obtain parameter estimates. These estimates are used to calculate predictions $\hat{y}(r, r')$ (cf. the procedure above), which satisfy restriction (6.15).
2. A multiplicative error term like in (6.10). Two estimation procedures may be followed. First, one may take logarithms and apply a non-linear estimation procedure. Secondly, one could take ratio's of observations, i.e. of the type $y(r, r')/y(r, r'')$, apply a logarithmic transformation, and use a linear least-squares approach. Predictions are generated as above.

Both procedures imply restrictions on the error term in order to satisfy condition (6.16). For example, in 1. we have

$$\sum_{r'} \epsilon(r, r') = 0 \qquad \text{for each } r \tag{6.19}$$

This makes the covariance matrix singular, so that an adaptation of the estimation procedure may be required (e.g. we could delete observations). In 2. the restrictions on the error term are not that simple and can best be ignored perhaps.

3. A different specification is the following:

$$y(r, r') = y(r, .)\, x_1(r)^{\alpha_{21}} z_1(r')^{\alpha_{31}} \exp\{-\alpha_4\, d(r, r')\}\, \epsilon(r. r') \Big/ \sum_{r'} x_1(r)^{\alpha_{21}} z_1(r')^{\alpha_{31}} \exp\{-\alpha_4 d(r, r')\}\epsilon(r, r') \quad (6.20)$$

A possible method is to take ratio's of observations and to apply a logarithmic transformation. Then linear least-squares can be used, and predictions obtained as above.

4. One minimises the sum of the squared residuals which satisfy the following relation:

$$y(r, r') = y(r, 1)\, x_1(r)^{\hat{\alpha}_{21}} z_1(r')^{\hat{\alpha}_{31}} \exp\{-\hat{\alpha}_4\, d(r, r')\}\, \hat{\epsilon}(r, r') \Big/ \sum_{r'} x_1(r)^{\hat{\alpha}_{21}} z_1(r')^{\hat{\alpha}_{31}} \exp\{-\hat{\alpha}_4\, d(r, r')\}\, \hat{\epsilon}(r, r') \quad (6.21)$$

with a hat indicating an estimate. These residuals can be chosen so as to sum up to zero for each r. The procedure of estimation starts with taking a logarithmic transformation. Then the transformed observations are again transformed to obtain deviations from the mean, for each r separately. For this transformed model linear least-squares may be used, and the estimates generate predictions along the line discussed before (cf. also Van Est and Van Setten, this volume).

Empirical applications of such constrained interaction models are mainly found in studies of trip distribution in transportation research and in spatial allocation models. Such applications do not pay much attention to the introduction of the error term, however. This may be induced by the fact that few attempts are undertaken to test the significance of the parameter-estimates.

From the discussion above it may be concluded that we prefer estimation procedures that incorporate the possibility of hypothesis testing. Since one would also like to see some tests on spatial autocorrelation and spatial heteroscedasticity, it seems that the M.L. approach presents the most flexible and constructive method, because it enables us to introduce several alternative hypotheses, which may be appropriately tested (given that the number of regions is not too small). Finally, in the introduction to this section we stated some stages in econometric analysis. These include also ex-ante validation of the model. For spatial interaction models such validation is rarely given. An exception is Openshaw (this volume) who did not find results that differ much from those of ex-post validation for deterministic independent variables.

2.6.4 *An extension of spatial interaction analysis*

In applied research the attention given to interaction models of the type discussed in the preceding sub-section is rather impressive. However, it may quite well be questioned whether the gravity hypothesis that forms the basis of these models, is indeed the most appropriate for explaining spatial flows. It is widely admitted that its foundation is rather weak, although artificial attempts have been done to justify the specification by means of entropy-maximising models. For suggestions on a more behavioural foundation of these models we may refer to Cesario and Smith (1975).

Given this state of knowledge, a valid approach consists of investigating a more extended class of interaction models which incorporates the multiplicative model as a special member. For the unconstrained model (6.8) we propose the following generalisation:

$$\tilde{y}(r, r') = \beta_1 + \sum_{k=1}^{K} \beta_{2k}\, \tilde{x}_k(r, r') + \epsilon(r, r') \qquad (6.22)$$

with a $\sim$ denoting a transformation of the variable (cf. below) and all independent variables indicated by $x_k\ (r, r'), k = 1, \ldots, K$ (these may depend on r, or r', or on both). For the error term the traditional assumptions apply.

For the transformation function we follow Box and Cox (1964) and define

$$\tilde{x}_k(r, r') = \frac{x(r, r')^{\beta_o} - 1}{\beta_o}, \qquad \beta_o \neq 0$$
$$= \ln x(r, r') \qquad \beta_o = 0 \tag{6.23}$$

with β_0 a transformation parameter. For other variables the transformation is defined similarly. We note that model (6.22) possesses the long-linear model (i.e. the multiplicative model) as a special case. Estimation of the parameters, including β_0, and significance testing may be done using a M.L. approach. In Bartels and Ter Welle (1977) we applied this specification to determine an optimal form for a migration relation explaining inter-provincial migration in the Netherlands. A similar approach in another context is used in Leech (1975), inter alia. Although this approach is tractable as far as parameter-estimation is concerned, difficulties arise if an unbiased predictor has to be designed for the un-transformed dependent variable. This problem seems to be ignored in the literature up till now so that little is known about the possibilities existing for this purpose.

Having discussed some topics related to single equation estimation with spatial data, we proceed to a brief introduction into some elements of stimultaneous models in a spatial context.

2.7 SIMULTANEOUS EQUATION MODELS WITH SPATIAL DATA

Our discussion of spatial models will be restricted to stochastic models, which require statistical techniques to obtain parameter-estimates. We shall give a brief introduction into some different classes of models.

A *first* class of models consists of models using only time-series data mainly for intra-regional variables for estimation purposes. They are purely analogous to the much better developed national models. The basic structure of the latter type of model is applied to a smaller area in space. An important impulse to this type of modelling was given by Klein (1969). Recent experience in the United States is summarised in Glickman (1974), (1977), and Klein and Glickman (1977).

Important features of this class of models are:

1. They are based on a top-down approach, i.e. national variables enter as exogenous in regional relations without any feedback. This implies that model-solutions will be inconsistent with the exogenous input.
2. Since only a limited number of short time-series is available for estimation, the number of variables included in the model will be small, its relationships simple (e.g. static) and small in number.
3. Explicit spatial elements and spatial interactions are virtually absent in these models.

Many models of this class have been estimated using single-equation OLS for the equations of the structural form to obtain parameter-estimates. Several simultaneous equations methods have been employed in Glickman (1976). Comparing different methods, he found the two-stage LS approach using principal components and an interative 2 SLS procedure using simulated values of the variables in the second step as performing better than iterated instrumental variables estimation for a specific model. In econometrics more studies have been done focusing on small-sample properties of simultaneous equations methods, which contain valuable information for application in the regional context.

A *second* class of models puts more emphasis on spatial characteristics. Examples are the Nova Scotia model (Czamanski, 1972) where inter-regional choice elements have been included by means of operational concepts like accessibility indices; the study of Bennett (1975) where a small linear model for a large number of areas is developed, which includes both spatial and temporal lags; the COMET model for the EEC countries (Barten et al., 1976) which contains country models with explicit linkages, and the LINK project (cf. Ball, ed., 1973) which aims at modelling the complete world using interrelated sub-models for different areas. If such models are being built for small areas one will frequently have to employ cross-section estimation, since time-series data will be absent in many cases. If short time-series are available, combined cross-section time-series estimation may be necessary. Glejser et al. (1973) used such a procedure to estimate a regional model for Belgium. It is obvious that the use of cross-section estimation makes investigation of parameter-stability required. For time-series, procedures based on recursive estimation have been used for this purpose, cf. Brown et al. (1975) and Bennett (1975). The latter author employs an adaptive parameter estimation procedure using a specific recursive least-squares approach as developed by Kalman. Study of extension of such approaches to cross-sectional data is strongly required (cf. also our discussion in Sub-section 6.2).

A *final* category of models that is very popular among regional scientists consists of regional input-output models. If exhausting sampling of firms is possible, these models are deterministic. However, in several cases limited surveys have to produce the information for constructing input-output tables and deriving input-output coefficients. In such a situation a stochastic approach to the determination of the coefficients is required, which could produce also standard errors of coefficient estimates. For estimating interregional trade patterns a spatial interaction model (cf. Section 6.3) can be used, cf. Snickars (this volume) for a further discussion of some estimation procedures in this case. In Gerking (1976) a 2 SLS procedure is followed to obtain consistent estimates and asymptotic standard errors for the input-output coefficients.

2.8 SOME REMAINING TOPICS

In this section we want to mention briefly some topics that seem to require more attention in futural spatial analysis.

The first concerns the *choice of spatial units*. Above we took the spatial configuration as a priori given. However, if data on a sufficient large number of small areas are available the possibility exists to design alternative sets of aggregated regions. This implies two choices,

- the *scale* of aggregation, i.e. the number or size of aggregated regions to be formed.
- the *composition* of the aggregated regions, i.e. which small areas are selected to form a new aggregated region.

It is well-known that the results of statistical analysis of spatial data do depend rather strongly on the scale of aggregation and the specific partition of the regional system. An illustration of this statement may be found in Openshaw (1977a, 1977b). Of course this dependency should not always be considered as a drawback, since we expect in spatial analysis a priori an influence of the spatial configuration of the phenomena to be studied at an aggregated level.

As noted before, the assumptions of several statistical methods may require the spatial data to posses certain characteristics, e.g. stationarity, homoscedasticity, which are not guaranteed by a specific spatial configuration. In this case transformation of the data may be in order to satisfy the required assumptions. Alternatively, a further specification of the covariance structure of the error terms may be attempted (cf. Streitberg, this volume). Hence, no uniform statement can be made about the treatment of the aggregation problem in spatial analysis. Instead we suggest investigating, in each case, whether the statistical assumptions are satisfied or not, and to apply some data transformations if necessary.

A different treatment of the problem of choosing a spatial configuration is given in Openshaw (1977a, 1977b). He proposes selecting that configuration which results in the optimal performance of a model in a spatial domain. Let us take as an example a spatial interaction model (cf. Section 6). Data are given for small spatial units. We aggregate these to new zones, and select that partition which yields a spatial interaction model with the best fit, e.g. measured by R^2. A tractable procedure for finding an approximate solution of this programming problem has been constructed by Openshaw. If the a priori constraints on the spatial delimination are small enough, this approach seems quite well defendable. In other situations, as when administrative boundaries suggest only a small number of spatial configurations, the approach as indicated above should be preferred.

A second topic that requires more attention is an attempt to account explicitly for errors in measurement in the data. In many cases the variables one likes to include on theoretical grounds will be unobservable, and one has to work with indicators or proxies of the unobservable variables. This will imply a measurement error. A general scheme for the treatment of errors-in-variables has been outlined by Jöreskog. He starts from the following system of linear structural equations (cf. Jöreskog, 1973):

$$\boldsymbol{A}\boldsymbol{a} = \boldsymbol{B}\boldsymbol{b} + \boldsymbol{u} \tag{8.1}$$

$$\boldsymbol{y} = \boldsymbol{a} + \boldsymbol{\mu} + \boldsymbol{v} \tag{8.2}$$

$$\boldsymbol{x} = \boldsymbol{b} + \boldsymbol{\gamma} + \boldsymbol{w} \tag{8.3}$$

with

$\boldsymbol{A}$ and $\boldsymbol{B}$ coefficient matrices

$\boldsymbol{a}$	a random vector of unobservable (latent) dependent variables
$\boldsymbol{b}$	a random vector of unobservable independent variables
$\boldsymbol{y}$	a random vector of observed indicators for the dependent variables.
$\boldsymbol{x}$	a random vector of observed indicators for the independent variables.
$\boldsymbol{u}, \boldsymbol{v}, \boldsymbol{w}$	vectors of disturbances

and

$$\boldsymbol{\mu} = \mathrm{E}(\boldsymbol{y})$$
$$\boldsymbol{\gamma} = \mathrm{E}(\boldsymbol{x})$$
$$\mathrm{E}(\boldsymbol{a}) = \mathrm{E}(\boldsymbol{b}) = \boldsymbol{0}$$

So, (8.1) describes the simultaneous relations between unobservable variables. The remaining equations indicate the correspondence between observed and unobservable variables. Criteria for identification and a maximum likelihood method for estimation are given in Jöreskog (1973). There it is also demonstrated that factor analysis (cf. Section 2.3.2) may be treated as a special case of this general approach. In the context of spatial data, this approach to errors in measurement seems to have received little or no attention.

A more empirically orientated procedure is followed by Openshaw (1976, and in this volume). He assumes the means of the observations to be known and simulates values from a normal distribution with a specified standard error, representing the degree of uncertainty in measurement. In this way the sensitivity of parameter-estimates towards errors in variables may be empirically established.

A third topic concerns the use of a priori ideas about the probability distribution of parameters during the estimation procedure. This is accomplished in Bayesian analysis. The principal ideas of this analysis and its use in spatial analysis are treated by Hepple in this volume, so that we suffice with referring to that contribution. The approach seems especially worthwhile to examine the robustness of models to departures from their assumptions.

2.9 FINAL REMARKS

Our summary of several statistical methods which are operational in the context of spatial data is far from complete. Discussion of other important statistical topics would extend the size of this paper too much, however. Some topics that have not received our attention are

- the use of logit models, e.g. in transport studies to determine the choice of a certain transport alternative (cf. Theil, 1971, for an introduction to logit models)
- the estimation of spatial diffusion models, cf. Cliff and Ord (1975)
- the introduction of stochastic elements in regional control models, cf. Bennett and Tan (this volume)

A general conclusion which may be drawn from our discussion is that several operational statistical methods are available for the analysis of spatial data, but still further developments into various directions are required to guarantee a satisfactory output in empirical applications.

REFERENCES

Abrahamse, A. P. J. and Koerts, J. (1971), 'New Estimators of Disturbances in Regression Analysis', *Journal of the American Statistical Association*, 66, 71–74.

Adelman, I. and Morris, C. T. (1965), 'A factor analysis of the interrelationship between social and political variables and per capita gross national product', *The Quarterly Journal of Economics*, 79, 555–578.

Adelman, I. and Morris, C. T. (1973), *Economic Growth and Social Equity in Developing Countries*, Stanford University Press, Stanford.

Anderson, T. W. (1958), *An introduction to multivariate analysis*, Wiley, New York.

Ball, R. (ed.), (1973), *The international linkage of national economic models*, North Holland, Amsterdam.

Barten, A. P., d'Alcantara, G. and Carrin, G. J. (1976), 'COMET. A medium-term macroeconomic model for the European Economic Community', *European Economic Review*, 7, 63-115.

Bartels, C. P. A. (1977a), *Economic aspects of regional welfare: income distribution and unemployment*, Nijhoff, Leiden.

Bartels, C. P. A. (1977b), 'The structure of regional unemployment in the Netherlands. An exploratory statistical analysis', *Regional Science and Urban Economics*, 7, 103–135.

Bartels, C. P. A. and Nijkamp, P. (1976), 'An empirical welfare approach to regional income distributions. Alternative specifications and estimates of income inequality measures', *Socio-Economic Planning Sciences*, 10, 117–128.

Bartels, C. P. A. and Bertens, A. M. (1976), 'A factor and regression analysis of regional differences in income-level and -concentration in the Netherlands', *Applied Economics*, 8, 179–192.

Bartels, C. P. A. and Hordijk, L. (1977), 'On the power of the generalized Moran contiguity coefficient in testing for spatial autocorrelation among regression disturbances', *Regional Science and Urban Economics*, 7, 83–101.

Bartels, C. P. A., Booleman, M. and Peters, W. H. (1977), 'The composition of and associations between regional unemployment series. An application of some time-series methods', *Research Memorandum 32, Faculty of Economics*, University of Groningen.

Barlett, M. S. (1964), 'The spectral analysis of two-dimensional point processes', *Biometrika*, 51, 299–311.

Bartlett, M. S. (1974), 'The statistical analysis of spatial pattern', *Advances in Applied Probability*, 6, 336-358.

Bartlett, M. S. (1975), *The statistical analysis of spatial pattern*, Chapman and Hall, London.

Barton, B. and Tobler, W. (1971), 'A spectral analysis of innovation diffusion', *Geographical Analysis*, 3, 182–186.

Bassett, K. and Tinline, R. (1970), 'Cross-spectral analysis of time-series and geographical research', *Area*, 1, 19–24.

Bassett, K. and Haggett, P. (1971), 'Towards short-term forecasting for cyclic behaviour in a regional system of cities', in M. Chisholm (ed.), *Regional Forecasting*, Butterworths, London, 389-413.

Batty, M. et al. (1973), 'Spatial system design and fast calibration of activity interaction-allocation models', *Regional Studies*, 7, 351–366.

Batty, M. and March, L. (1976), 'The method of residues in urban modelling', *Environment and Planning*, A, 8, 189–214.

Bennett, R. J. (1974), 'Process identification for time series modelling in urban and regional planning', *Regional Studies*, 8, 157-174.

Bennett, R. J. (1975a), 'The representation and identification of spatio-temporal systems: an example of population diffusion in North-West England', *Transactions of the Institute of British Geographers*, 66, 73-94.

Bennett, R. J. (1975b), 'Dynamic systems modelling of the North-West region', *Environment and Planning*, A, 7, I: Spatio-temporal representation and identification, 525-538, II: Estimation of the spatio-temporal policy model, 539–566, III: Adaptive-parameter policy model, 617–636, IV: Adaptive spatio-temporal forecast, 887–898.

Bennett, R. J. and Haining, R. P. (1976), 'Space-time models: an introduction and concepts', *Occasional Paper No. 28*, Dept. of Geography, University College, London.

Berry, B. J. L. (1971), 'Problems of Data Organization and Analytical Methods in Geography', *Journal of the American Statistical Association*, 66, 510-523.

Besag, J. E. (1974), 'Spatial interaction and the statistical analysis of lattice systems', *Journal of the Royal Statistical Society*, series B, 36, 192–236.

Bodson, D. and Peeters, D. (1975), 'Estimation of the coefficients of a linear regression in the presence of spatial autocorrelation: An application to a Belgian labour-demand function', *Environment and Planning*, A, 7, 455-472.

Boyce, D. E., Fahri, A. and Weischedel, R. (1974), *Optimal subset selection, multiple regression, interdependence and optimal network algorithms*, Springer Verlag, Berlin.

Box, G. E. P. and Cox, D. R. (1964), 'An analysis of transformations', *Journal of the Royal Statistical Society*, Series B, 26, 211–243.

Bradu, D. and Mundlak, Y. (1970), 'Estimation in lognormal linear models', *Journal of the American Statistical Association*, 65, 198–211.

Brillinger, D. R. (1975), *Time series. Data Analysis and Theory*, Holt, Rinehart and Winston, New York.

Brown, R. L., Durbin, J. and Evans, J. M. (1975), 'Techniques for testing the constancy of regression relationships over time', *Journal of the Royal Statistical Society*, Series B, 37, 149-163.

Cesario, F. J. (1975), 'Linear and non-linear regression models of spatial interaction', *Economic Geography*, 51, 69–77.

Cesario, F. J. and Smith, T. E. (1975), 'Directions for future research in spatial interaction models', *Papers R.S.A.*, 35, 57–72.

Chatterji, M. (1976), 'Some suggested approaches to spatial regularities', in M. Chatterji (ed.), *Space location and regional development*, Pion, London.

Cheng, D. C. and Iglarsh, H. J. (1976), 'Principal component estimators in regression analysis', *Review of Economics and Statistics*, 58, 229–234.

Cliff, A. D. and Ord, J. K. (1969), 'The problem of spatial autocorrelation', In A. J. Scott (ed.), *Studies in Regional Science*, Pion, London, 25-55.

Cliff, A. D. and Ord, J. K. (1970), 'Spatial autocorrelation: a review of existing and new measures with applications', *Economic Geography*, 46, 269-292.

Cliff, A. D. and Ord, J. K. (1972), 'Testing for spatial autocorrelation among regression residuals', *Geographical Analysis*, 6, 267-284.

Cliff, A. D. and Ord, J. K. (1973), *Spatial Autocorrelation*, Pion, London.

Cliff, A. D. and Ord, J. K. (1975), 'Model building and the analysis of spatial pattern in human geography', *Journal of the Royal Statistical Society*, 37, 297–328.

Cliff, A. D., Haggett, P., Ord, J. K., Bassett, K. and Davies, R. (1975), *Elements of spatial structure*, Cambridge University Press, Cambridge.

Czamanski, S. (1969), 'Regional Econometric Models: A Case Study of Nova Scotia', In A. J. Scott (ed.), *Studies in Regional Science*, Pion, London, 143–180.

Czamanski, S. (1972), *Regional Science Techniques in Practice: The Case of Nova Scotia*, Heath, Lexington.

Dacey, M. F. (1968), 'A review on measures of contiguity for two and k-color maps', In B. J. L. Berry and D. R. Marble (eds.), *Spatial Analysis. A reader in statistical geography*, Prentice Hall, New Jersey, 479-495.

Dhrymes, P. J. (1970), *Econometrics, statistical foundations and applications*, Harper and Row, New York.

Dubbelman, C., Abrahamse, A. P. J. and Koerts, J. (1972), 'A new class of disturbance estimators in the general linear model', *Statistica Neerlandica*, 26, 128-141.

Fisher, W. D. (1971), 'Econometric estimation with spatial dependence', *Regional and Urban Economics*, 1, 19–40.

Fishman, G. S. (1969), *Spectral Methods in Econometrics*, Harvard University Press, Cambridge.

Geary, R. C. (1954), 'The contiguity ratio and statistical mapping', *The Incorporated Statistician*, 5, 115–141.

Gerking, S. D. (1976), *Estimation of stochastic input-output models*, Nijhoff, Leiden.

Glejser, H., Van Daele, G. and Lambrecht, M. (1973), 'First experiments with an econometric regional model of the Belgian economy', *Regional and Urban Economics*, 3, 301–314.

Glickman, N. J. (1974), 'Son of the specification of regional models', *Papers and Proceedings of the Regional Science Association*, 32, 155–177.

Glickman, N. J. (1976), 'A note on simultaneous equation estimation techniques applications with a regional econometric model', *Regional Science and Urban Economics*, 6, 275–287.

Glickman, N. J. (1977), *Econometric analysis of regional systems. Explorations in model building and policy analysis*, Academic Press, New York.

Goldberger, A. S. (1968), 'The interpretation and estimation of Cobb-Douglas functions', *Econometrica*, 35, 464–472.

Goldberger, A. S. (1972), 'Structural equation methods in the social sciences', *Econometrica*, 40, 979–1001.

Goldfeld, S. M. and Quandt, R. E. (1972), *Non-linear Methods in Econometrics*, North Holland, Amsterdam.

Granger, C. W. J. (1969), 'Spatial data an time-series analysis', In A. J. Scott (ed.), *Studies in regional science*, Pion, London, 1–24.

Granger, C. W. J. and Hatanaka, M. (1964), *Spectral Analysis of Economic Time Series*, Princeton University Press, Princeton.

Hampton, P. and Rayner, A. C. (1977), 'The use of multivariate methods in economics with reference to regional analysis', *Regional Science and Urban Economics*, 7, 267–287.

Harkness, J. P. (1968), 'A spectral-analytical test of the long-swing hypothesis in Canada', *The Review of Economics and Statistics*, 50, 429-436.

Harvey, A. C. (1976), 'Estimating regression models with multiplicative heteroscedasticity', *Econometrica*, 44, 461–465.

Harvey, A. C. and Phillips, G. D. A. (1974), 'A comparison of the power of some tests for heteroscedasticity in the general linear model', *Journal of Econometrics*, 2, 307–316.

Heien, D. M. (1968), 'A note on log-linear regression', *Journal of the American Statistical Association*, 63, 1034–1038.

Hepple, L. W. (1974), 'The impact of stochastic process theory upon spatial analysis in human geography', In C. Board, R. J. Chorley, P. Haggett, D. R. Stoddard (eds.), *Progress in Geography, Volume 6, International Reviews of Current Research,* Edward Arnold, London, 89–142.

Hepple, L.W. (1975), 'Spectral techniques and the study of interregional economic cycles', In R. Peel, M. Chisholm and P. Haggett (eds.), *Processes in Physical and Human Geography*, Bristol Essays, 392–408.

Hepple, L. W. (1976), 'A maximum likelihood model for econometric estimation with spatial series', In I. Masser (ed.), *Theory and Practice in Regional Science*, London Papers in Regional Science, 6, Pion, London.

Hordijk, L. (1974), 'Spatial correlation in the disturbances of a linear interregional model', *Regional and Urban Economics*, 4, 117–140.

Hordijk, L. and Nijkamp, P. (1976), 'Dynamic models of spatial autocorrelation', *Paper European Meeting of the Econometric Society*, Helsinki.

Hordijk, L. and Paelinck, J. (1976), 'Some principles and results in spatial econometrics', *Paper Netherlands Economc Institute*, Rotterdam.

Huijbregts, C. J. (1975), 'Regionalized variables and quantitative analysis of spatial data', In J. C. Davis and M. J. McCullagh (eds.), *Display and analysis of spatial data*, Wiley, London.

Jaumotte, Ch. and Paelinck, J. H. P. (1971), 'The differential economic structures of the Belgian provinces: a time varying factor analysis', *Regional and Urban Economics*, 1, 41–75.

Jenkins, G. M. and Watts, D. G. (1968), *Spectral analysis and its applications*, Holden-Day, Amsterdam.

Johnson, K. H. and Lyon, H. L. (1973), 'Experimental evidence on combining cross-section and time-series information', *Review of Economics and Statistics*, 55, 465–474.

Johnson, S. R., Reimer, S. C. and Rothrock, T. P. (1973), 'Principal components and the problem of multicollinearity', *Metro-economica*, 25, 306–317.

Johnston, R. J. (1976), 'On regression coefficients in comparative studies of the "frictions of distance"', *Tijdschrift voor Econ. en Soc. Geografie*, 67, 15–28.

Jöreskog, K. G. (1973), 'A general method for estimating a linear structural equation system', In A. S. Goldberger and O. D. Ducan (eds.), *Structural Equation Models in the Social Sciences*, Seminar Press, New York.

Kau, J. B. and Lee, Ch. F. (1977), 'A random coefficient model to estimate a stochastic density gradient', *Regional Science anq Urban Economics*, 7, 169–177.

Kendall, M. G. (1973), *Time-series*, Griffin, London.

King, L., Casetti, E. and Jeffrey, D. (1969), 'Econometric Impulses in a Regional System of Cities: A Study of Spatial Interaction', *Regional Studies*, 3, 213–218.

Klein, L. R. (1969), 'The specification of regional econometric models', *Papers of the Regional Science Association*, 25, 105–115.

Klein, L. R. and Glickman, N. R. (1977), 'Econometric model-building at regional level', *Regional Science and Urban Economics*, 7, 3–23.

Koerts, J. and Abrahamse, A. P. J. (1971), *On the theory and application of the general linear model*, Rotterdam University Press, Rotterdam.

Kooijman, S. A. L. M. (1976), 'Some remarks on the statistical analysis of grids especially with respect to ecology', *Annals of systems research*, 5.

Kuh, E. (1959), 'The validity of cross-sectionally estimated behavior equations in time series applications', *Econometrica*, 27, 197–214.

Lebart, L. (1969), 'Analyse statistique de la contiguité', *Publications de l'Institut de Statistique de l'Université de Paris*, 18, 81–112.

Leech, D. (1975), 'Testing the error specification in non-linear regression', *Econometrica*, 43, 719–725.

Malinvaud, E. (1966), *Statistical methods of econometrics*, North Holland, Amsterdam.

Marquardt, D. W. and Snee, R. N. (1975), 'Ridge regression in practice', *The American Statistician*, 29, 3–20.

Martin, R. L. (1974), 'On spatial dependence, bias and the use of first spatial differences in regression analysis', *Area*, 6, 185–194.
Masser, I. (1977), 'A comparitive analysis of spatial representation in doubly constrained interaction models', *Environment and Planning* A, 9, 759–769.
McCallum, B. T. (1970), 'Artificial orthogonalization in regression analysis', *The Review of Economics and Statistics*, 52, 110-113.
McCamley, F. (1973), 'Testing for spatially autocorrelated disturbances with application to relationships estimated using Missouri county data', *Regional Science Perspectives*, 3, 89–104.
Merkies, A. H. Q. M. (1975), 'A constructive test on heteroscedasticity', *Research Memorandum 19*, Free University, Amsterdam.
Mizon, G. E. (1977), 'Inferential procedures in non-linear models: an application in a UK industrial cross section study of factor substitution and returns to scale', *Econometrica*, 45, 1221–1242.
Moran, P. A. P. (1950), 'Notes on continuous stochastic phenomena', *Biometrika*, 37, 17–23.
Mulaik, S. A. (1972), *The foundations of factor analysis*, McGraw-Hill, New York.
Nerlove, M. (1971), 'Further evidence on the estimation of dynamic economic relations from a time series of cross sections', *Econometrica*, 39, 359–382.
Nijkamp, P. (1976), 'Quality of life and changes in human settlement patterns', *Research Memorandum 63*, Free University, Amsterdam.
Ord, J. K. (1975), 'Estimation methods for models of spatial interaction', *Journal of the American Statistical Association*, 70, 120–126.
Openshaw, S. (1976), 'An empirical study of some spatial interaction models', *Environment and Planning A*, 8, 23–41.
Openshaw, S. (1977[a]), 'A geographical solution to scale and aggregation problems in region-building, partitioning, and spatial modelling', *Transactions of the Institute of British Geographers*, New Series, 459–472.
Openshaw, S. (1977[b]), 'Optimal zoning systems for spatial interaction models', *Environment and Planning A*, 9, 169–184.
Paelinck, J. H. P. and Wagenaar, Sj. (1976), 'Treatment of multidimensional data: a review and some contributions', *Paper Netherlands Economic Institute*, Rotterdam.
Phillips, G. D. A. and Harvey, A. C. (1974), 'A Simple Test for Serial Correlation in Regression Analysis', *Journal of the American Statistical Association*, 69, 935–939.
Ranner, A. P. (1974), 'Analyzing relations between regions and their surroundings', A canonical correlation approach', *Regional and Urban Economics*, 4, 141–162.
Rayner, J. N. and Colledge, R. G. (1972), 'Spectral analysis of settlement patterns in diverse physical and economic environments', *Environment and Planning*, 4, 347–371.
Rayner, J. N. and Colledge, R. G. (1973), 'The spectrum of U.S. route 40 Reexamined', *Geographical Analysis*, 5, 338–350.
Rice, J. R. and White, J. S. (1964), 'Norms for smoothing and estimation', *SIAM Review*, 6, 243–255.
Ripley, B. D. (1977), 'Modelling spatial patterns (with Discussion)', *Journal of the Royal Statistical Society*, Series B, 39, 172–212.
Ronner, A. E. (1977), *P-norm estimators in a linear regression model*, dissertation, University of Groningen.
Scott, J. T. (1966), 'Factor analysis and regression', *Econometrica*, 34, 552–562.
Smith, D. M. (1973), *The geography of social well-being in the United States. An introduction to territorial social indicators*, McGraw-Hill, New York.
Teekens, R. and Koerts, J. (1972), 'Some statistical implications of the log transformation of multiplicative models', *Econometrica*, 40, 793–819.
Theil, H. (1965), 'The analysis of disturbances in regression analysis', *Journal of the American Statistical Association*, 60, 1067–1079.
Theil, H. (1971), *Principles of econometrics*, Wiley, New York.
Thomann, R. V. (1974), 'Some statistical aspects of water quality in the Delaware river', In J. W. Pratt (ed.), *Statistical and mathematical aspects of pollution problems*, Dekker, New York, 279–311.
Tobler, W. R. (1969), 'The spectrum of U.S. 40', *Papers of the Regional Science Association*, 23, 45–52.

Tobler, W. R. (1973), 'Regional analysis time series extended to two dimensions', *Geographia Polonia*, 25, 103–106.

Tyson, P. D. (1971), 'Spatial variation of rainfall spectra in South Africa', *Annals of the Association of American Geographers*, 61, 711–720.

Sen, A. (1976), 'Large sample-size distribution of statistics used in testing for spatial correlation', *Geographical Analysis*, 9, 175–184.

Van der Knaap, G. A. and Lesuis, P. J. J. (1976), 'The integrated analysis of spatial systems: a problem area. An application to the Dutch provinces', *Inst. for Economic Research, discussion paper 7601/G*, Erasmus University, Rotterdam.

Van Est, J. and Van Setten, J. (1977), 'Calibration methods: application for some spatial distribution models', *Working Paper 6, TNO*, Delft.

Van Rompuy, P. and Tejano, A. (1975), 'Intertemporal factor analysis applied to Belgian standard of living indicators', *Regional Science Research Paper, nr. 5*, Centrum voor Economische Studiën, Katholieke Universiteit, Leuven.

Vinod, H. D. (1976), 'Canonical ridge and econometrics of joint production', *Journal of Econometrics*, 4, 147–166.

Wilson, A. G. (1974), *Urban and regional models in geography and planning*, Wiley, London.

Part 2: Exploratory statistical analysis

3. The analysis of geographical maps

BRIAN D. RIPLEY*

3.1 INTRODUCTION

Much spatial data is in the form of maps of isolated objects such as towns, forts and stores in geography and archaeology, plants and nests in ecology, stars in astronomy and many other examples from a wide variety of disciplines. Three examples of such maps are given in figures 1, 4 and 9 (see appendix to this chapter). We concentrate on methods of analysis which aim to detect interaction between the objects exhibited by regularity or clustering of their spatial pattern. The traditional methods are reviewed and compared with some recent developments.

We assume that our raw spatial data is a map (or a list of co-ordinates) giving the locations of N 'small' objects within a known region E of measured area A. For example, figure 4 shows the locations of 69 towns within a 40 mile square in Spain. (It is often convenient to take our unit of length to be the side of such a square; then $A = 1$). We will consider only two-dimensional maps, although the methods described here can easily be generalised to three dimensions or to objects over the whole surface of the earth (e.g. earthquakes).

Assumptions of homogeneity and isotropy are implicit or explicit in virtually all the methods to be discussed. Isotropic patterns are taken to be those with no preferred direction. For the underlying process which produced the observed map to be homogeneous we need to assume that the whole of E is equally habitable and that the interactions depend only on the relative positions of the objects. These are restrictive assumptions, often questioned in geography, yet without them we are as yet able to perform only the most

*Department of Mathematics, Imperial College, London SW7, U.K.

superficial analysis of interaction. It should always be borne in mind that apparent clustering, for example, could be caused by inhomogeneity of the region rather than by interaction. One way to minimize this problem is to intensively study small sections of maps representing the subregions which are considered to be homogeneous.

3.2 METHODS OF ANALYSIS

Probably the oldest method for the detection of interaction in maps is quadrat sampling which has been much used in ecology. In its original form the placing of the sampling squares was intended to be random, but the method has also been used with the systematic division of the mapped region into squares. Geographical examples, including one re-analysed below, are given by Rogers (1974). As only the numbers of squares containing 0, 1, 2, . . . objects are retained, this method virtually destroys the spatial nature of the data. By a judicious choice of the size of the sampling unit it can yield useful information, but it does seem undesirable to waste so much of the available information.

Recognising the need to combine data from the many basic sampling units, Greig-Smith and later ecologists (Greig-Smith, 1952, 1964; Kershaw, 1957; Pielou, 1977; Usher, 1969, 1975; Errington, 1973; Hill, 1973; Mead, 1974) have devised analyses which combine the counts from neighbouring basic squares. These procedures are essentially tests of the equality of the expected counts within subregions of various specified sizes. Perhaps the most sophisticated of these tests is that derived by Zahl (1974). Other ecologists have criticized such procedures, preferring the correlation between the numbers in pairs of quadrats at varying distances apart (Goodall, 1974; Pielou, 1977), i.e. a spatial autocorrelation approach (Cliff and Ord, 1973). Spatial spectra of the grid of counts or of the objects themselves have also been considered (Bartlett, 1964, 1975; Ford, 1976).

The third traditional approach is based on measurement of the distances from an object to its nearest neighbour, or from randomly located points to the nearest object. Most such 'distance methods' (also referred to in ecology as 'plotless' sampling) were designed for use in preliminary fieldwork, to estimate the number of points per unit area or to test for randomness (defined in section 3.4). Their sampling distributions were derived assuming that the sampled points and objects are far apart, in particular, not more than one tenth of the objects being involved in the sample. We will refer to this as *sparse* sampling. However, Clark and Evans (1954) and Brown (1975) have proposed tests which are based on all the distances from each object to its nearest neighbour. Let these distances be $(d_1, \ldots, d_N)$. Bartlett (1975) used the index of Eberhardt (1967), $N\Sigma d_i^2 / (\Sigma d_i)^2$. Clark and Evans referred the average $\bar{d}$ of these distances to a normal distribution with mean and variance calculated for sparse sampling. Unfortunately these values are often considerably in error; recent corrections are given below, in section 3.4.

In Ripley (1976, 1977) I started from the idea that usually a single number, as given by distance methods, is not a sufficient summary of the data. Indeed the Greig-Smith approach produces a graphical summary. There I concentrated on finding stochastic models which could plausibly have produced the map. Thereby the fit was assessed by two cumulative distribution functions, roughly of the distances between pairs of objects and the distance from a randomly chosen point to the nearest object. The use of the first of these distribution functions was suggested by Bartlett (1964). These two graphical summaries still do not characterize the pattern completely; they were chosen because they are related to simple properties of the models and have a simple interpretation.

3.3 MODELS

Although these statistics could be considered merely as summaries of the map, their interpretation is greatly helped by considering some simple mechanisms which might have produced the map. As we are not interested in explaining precisely why the objects are where they are, but rather in exploring their arrangement relative to each other, it is natural to consider stochastic processes.

The most commonly quoted process is a 'random' pattern, one with no interaction between the objects. As usual in statistics its pre-eminence is not because it is often true (no precise hypothesis is ever true) but that it provides an extremely convenient datum from which to assess interactions.

In fact there are two subtly different definitions of a random process. In the first approach the number N of points is assumed to be fixed, so the locations of the points are N random two-dimensioal vectors. Our assumption of homogeneity entails that each is separately uniformly distributed over E. The 'random' process, called the binomial process by Elfing (1947), assumes that these N vectors are independent. Alternative processes will allow dependence on the relative positions. Note that we cannot discuss invariance of such a process under translations or rotations since these will rarely map E onto itself. Indeed the restriction to a fixed number of points in E, but not in subregions shows the privileged position of this region.

The alternative view suggests that E is just part of some larger region within which the process is operating, so the number of points in E should be variable. (This emphasizes that E should ideally be chosen without reference to the objects themselves. This is rarely so, since a homogeneous region is usually chosen by inspection, but increasing E to include a particular point is definitely cheating!) We now suppose that the number of objects in E is a Poisson random variable with mean A, the area of E. Thus the binomial process is this Poisson point process conditioned on the total number of points in E. The Poisson distribution is necessary for the numbers of points in non-overlapping subregions to be independent, an indication of the absence of interaction. Further if we suppose that we have observed the part within E of a Poisson process operating in a larger region, the observed process will again be a Poisson process. (This is not so for a binomial process.) If E were an interval of time we would have defined the well-known Poisson process.

Since we now conceive of the process extending beyond E it is possible to insist that it is invariant under both translations and rotations, i.e. that the joint distribution of the numbers of objects in a number of subregions depends only on the relative positions of these subregions. The random variables $Z(F)$ denoting the number of objects in subregions F are important. It follows from the assumed invariance that $E(Z(F))$ depends only on the area of F, so $E(Z(F)) = \lambda$ area (F) where λ is a constant known as the *intensity*, the mean number of points per unit area. We can also summarize the covariances between $Z(F_1)$ and $Z(F_2)$ for two subregions F_1 and F_2. In Ripley (1976, 1977) I showed that these covariances could be calculated by an unilluminating formula from a non-negative increasing function $K(t)$. We can interpret $\lambda K(t)$ as the expected number of further objects within distance t of an arbitrary object. With some technical conditions $g(s) = (\lambda^2 dK/ds)/2\pi s$

can be interpreted as the joint density for the occurrence of two points distance s apart. Thus if F_1 and F_2 are small subregions distance s apart,

$$pr\,(\text{one object in } F_1,\text{ another in } F_2) = g(s)\,\text{area}\,(F_1)\,\text{area}\,(F_2)$$

where the equality is approximate. (The factor $2\pi s$ arises as the area of the annular region of points between s and $s+ds$ from an object, divided by ds.) We can specify a process by giving

$$pr\,(\text{no points in } F_1, \ldots, F_n)$$

for all finite collections of circular subregions (Ripley, 1977). As this function is specified by the radii of the discs and the relative positions of their centres, it is a complex function. We will consider merely the case of one disc, so considering $p(t) = pr(Z(F) > 0)$ for any circular region F of radius t.

The parameters λ, $K(t)$ and $p(t)$ do *not* completely specify the model. They do however summarize many of its important features.

For a Poisson process $p(t) = 1 - \exp\{-\lambda\pi t^2\}$, the probability that a Poisson variable of mean λ times the area of a disc of radius t is not zero. By the independence of the counts in non-overlapping subregions $g(s) = \lambda^2$, so

$$K(t) = \int_0^t g(s)\,2\pi s\,ds/\lambda^2 = \pi t^2.$$

These definitions depend essentially on the invariance of the process and so cannot be properly applied to the binomial process. Nevertheless because this process lacks edge effects (see below) we find[1]

$$g(s) = N(N-1)/A^2, K(t) = (N-1)\pi t^2/N$$
$$\text{and } p(t) = 1-(1-\pi t^2/A)^N.$$

1. To find $g(s)$ we choose one of the $N(N-1)$ pairs of objects. The two objects are independent, so

$$pr(\text{first object in } F_1,\text{ second in } F_2) = \text{area}\,(F_1)\,\text{area}\,(F_2)/A^2.$$

Summing over the pairs we find $g(s) = N(N-1)A^{-2}$. Since $\lambda = N/A$, $K(t) = N(N-1)\pi t^2/A^2(N/A)^2$. The probability that a given object falls in a disc of radius t is $\pi t^2/A$, so the probability that none of the objects falls in the disc is $(1-\pi t^2/A)^N$ and this is $1-p(t)$.

We will need two further types of models. Neyman-Scott cluster processes are defined in two stages. First parent centres are placed following a Poisson process of intensity α. Then a random number n of daughters is placed independently around each parent. The daughter processes are independent of each other and identically distributed apart from their centring. For such a process let F be the distribution function of the distance between any two daughters with the same parent. Then

$$K(t) = \pi t^2 + \alpha E(n(n-1))F(t)/\lambda^2$$

Here the first term refers to pairs of points with different parents, the second to pairs of siblings.[2] Notice that $K(t)$ always exceeds that of a Poisson process with the same intensity, πt^2; by an increasing amount. Thus large values of K may be interpreted as clustering on a scale determined from the values of t for which this difference increases rapidly. It *is* possible to calculate p for this model, but the resulting formula is forbidding.

We also need models to express repulsion or exclusion between objects. (For instance we would not expect cities to occur very close together. We would expect K to be below and p above the same function for a Poisson process of the same intensity, for K measures the propensity of points to be close together and p the uniformity of the pattern. We revert to our first approach and modify the binomial process. Suppose we wish to model the packing of discs of radius R which are not allowed to overlap. Then their centres form a process in which points are not allowed to be less than $2R$ apart. A simple model, termed an SSI process by Diggle et al. (1976) is defined as follows. Place the first point uniformly

---

2. Fix an arbitrary object; $\lambda K(t)$ is the expected number of further objects within distance t. This is the sum of objects in different clusters, $\lambda\pi t^2$ as in a random process by the independence of the clusters, and objects in the same cluster. Let p_n be the probability that a cluster has n daughters, and suppose the fixed object is in a cluster of n objects. The contribution to $\lambda K(t)$ from the $(n-1)$ other points in the cluster is $(n-1)F(t)$. Since we choose an arbitrary *object* the probability of a cluster of size n is $np_n/E(n)$, so

$$\lambda K(t) = \lambda\pi t^2 + \sum_{n=1}^{\infty} (n-1)F(t)np_n/E(n)$$

$$= \lambda\pi t^2 + E(n(n-1))F(t)/E(n).$$

The result follows from $\lambda = \alpha E(n)$. This particular deviation is due to Peter Diggle.

within the region E, then place each subsequent point uniformly over that part of the region where it would not be nearer than $2R$ to any existing point. An alternative, different, approach is to condition that binomial process on the event that no pair of points within E are less than $2R$ apart. This is a special case of the Strauss process (Strauss, 1975, corrected by Kelly and Ripley, 1976) in which the joint density of the N points is proportional to $c^{s(r)}$ where c in a non-negative constant and $s(r)$ is the number of pairs of points closer than r. If we take $c = 0, r = 2R$ we obtain the process defined above. If N is allowed to vary (but we still define the process only on E) we can characterise this process by the probability density of an object being found at a point depending only on the number of objects within r of that point.

It is fairly intuitive that for these processes $E(Z(F))$ depends not only on the area of F but also on its location, particularly whether it is near the edge of E. Thus we cannot define K and p without removing this edge effect. Usually E will be a square. Then we may regard E as a torus and define the distance between (x_1, y_1) and (x_2, y_2) as

$$\sqrt{(a(x_1, x_2)^2 + a(y_1, y_2)^2)}$$

where

$$a(x_1, x_2) = \min(|x_1 - x_2|, s - |x_1 - x_2|) \text{ and}$$

s is the length of a side of the square.

This makes points on opposite edges of the square near as they would be if we glued together opposite sides of the square to make a torus. With this definition of distance the processes on E defined in the last paragraph lose their edge effects. We can then define K and p but it is still not possible to calculate them explicitly.

Many other models have been defined at least heuristically; some are surveyed in Hodder and Orton (1976) and Ripley (1977). Defining models for maps is quite a tricky operation since there are many ways to consider subsets of the class of objects and it is easy to inadvertently assign two different probabilities to one event. In exploratory work it is probably best to stick to simple models such as those given here. Ultimately it would be desirable to fit a model defined by some theory to explain the pattern, such as a central-place or gravity model.

In the sequel, estimators of K and p are used to assess the fit of these models to data.

3.4 TESTS FOR RANDOMNESS

We return to the analysis of a given map. Our first null hypothesis is that it was generated by a homogeneous process without interaction, i.e. we assume a binomial or Poisson process. Later we may wish to test the fit of some other model. Incidentally, a process without interaction invariant under translations is automatically isotropic; the detection of directionality is a virtually unexplored problem.

We consider first the distance methods introduced in section 3.2. As there, let $d_1, \ldots, d_N$ be the distances from each object to its nearest neighbour. Let $E(d)$ and var (d) denote the mean and variance of any of these distances, and $\bar{d}$ their average. Clark and Evans (1954) defined two statistics:

$$CE = (\bar{d} - E(d))/\sqrt{(\mathrm{var}(d)/N)}$$

$$R = \bar{d}/E(d).$$

They assert that for a random process CE has approximately a standard $N(0, 1)$ distribution, and values of R larger than one indicate regularity whereas values smaller than one represent clustering. They used the sparse sampling formulae

$$E(d) = 0.5\sqrt{(A/N)}, \qquad \mathrm{var}(d) = 0.0683\, A/N.$$

There are two errors in the use of these formulae. First whereas $E(\bar{d}) = E(d)$, $\mathrm{var}(\bar{d}) \neq \mathrm{var}(d)/N$ since the distances $d_1, \ldots, d_N$ will be slightly correlated even for a random process (Diggle, 1976); finding d_i involves considering the distances to all the other points. Thus we should consider

$$CCE = (\bar{d} - E(d))/\sqrt{\mathrm{var}(\bar{d})}.$$

Much more serious is the assumption in the sampling formulae that the points are not near the edge of the region. Matérn (1972) showed that for a binomial process on E, where E is a 'compact region' of area A and perimeter length P,

$$E(d) = \sqrt{(A/(4N-1))} + 0.051\, P/N,$$

$$\mathrm{var}(d) = 0.0683\, A/N + 0.0357\, P\sqrt{A}\, N^{-3/2}.$$

Recently Donnelly (1978) has shown that with the following formulae *CCE* is very close to $N(0, 1)$–distributed on the hypothesis of a binomial process:

$$E(d) = 0.5\sqrt{(A/N)} + (0.514 + 0.042/\sqrt{N})P/N,$$

$$\mathrm{var}(\bar{d}) = 0.07\, A/N^2 + 0.04\, P\sqrt{A}\, N^{-5/2}.$$

In the sequel we will use *CE* with the original formulae, *CCE* with Donnelly's formulae, and sometimes *CCE* for distances on a torus, for which $P = 0$.

Brown (1975) considered two other statistics based on all the object to nearest neighbour distances. These are the coefficient of variation of the squared distances

$$B_1 = \Sigma(d_i^2 - d^*)^2/(N-i)(\ d^*))^2$$

$$\text{where } d^* = N^{-1}\ \Sigma d_i^2,$$

and the ratio of the geometric to arithmetic means of the squared distances

$$B_2 = (\prod_{i=1}^{N} d_i^2)^{1/N} / d^*.$$

The sampling distributions for a Poisson or binomial process are not known exactly. For *sparse* sampling from a Poisson process, $M = -2N \ln B_2$ has a known distribution – that of Bartlett's test of homogeneity of variances (used by Pollard, 1971; Diggle, 1977, as a test of heterogeneity), so $6M/7$ is approximately distributed as a chi-squared statistic with $N-1$ degrees of freedom. But in the present situation this can only be a rough guide. Even the sparse sampling distribution of B_1 is unknown. One advantage of Brown's statistics is that they do not involve the area of E. This is very relevant to their original application to nests of birds, e.g. Newton et al. (1977) in which the process is confined to a very irregular disconnected piece of suitable habitat. Of course the sampling distribution depends on the *shape* of the region E.

We do not know the distribution of these test statistics under our null hypothesis, so we may use simulation to approximate to this distribution. Since we are interested in the significance test of the goodness-of-fit of a process, the most efficient way to do this is to use the Monte-Carlo test procedure suggested by Barnard in the discussion of Bartlett (1963) and used in a spatial context by Cliff and Ord (1973), Ripley (1977) and Besag and Diggle (1977). Suppose we have the value of a test statistic X. We simulate our hypothesized process m times, generating values $X_1, \ldots, X_m$. Then if the null hypothesis is true we have $m + 1$ samples from the same distribution, so the probabilities that X is the largest or smallest are each $1/(m + 1)$. More generally we can rank the values. If there are r values as extreme as X (in a one-sided or two-sided sense) then the null hypothesis is rejected by a test of exact significance level $r/(m + 1)$. Thus $m = 19$ or 39 are suitable choices for one- and two-sided 5% tests. Obviously the value of m affects the power of the test against specified alternatives, since larger values of m will allow us to find the true significance point more accurately. The examples here were based on $m = 99$. Note that we can use the most appealing test statistic, and that we could test the fit of any model which we can simulate.

My tests are based on estimators of K and p. Let

$$\hat{K}(t) = \Sigma k(x, y)/\lambda^2 A,$$

the sum being over *ordered* pairs of objects x and y with $0 < d(x, y) < t$. Here $1/k(x, y)$ is the proportion within E of the circumference of the circle centered on x passing through y. Thus $\hat{K}(t)$ is a weighted sum of pairs of objects less than t apart. The point of the weights is to make $\hat{K}(t)$ an unbiased estimator of $K(t)$ *for any process* provided t is less than the circumradius of E (Ripley, 1976). Of course λ is rarely known. In Ripley (1977) I estimated λ^2 by $\hat{\lambda}^2 = (N/A)^2$. It might be better to use the unbiased estimator of λ^2, $N(N - 1)A^{-2}$. With this choice $\hat{K}(t)$ is again unbiased for a Poisson process (although for a binomial process we should use $\hat{\lambda}^2$ for unbiasedness!). For other processes $\hat{K}(t)$ is negligibly biased; the advantage of a nearly unbiased estimator is that one can build up one's intuition about its mean without taking account of the shape of E. This is not so for the unweighted empirical interpoint distance distribution function $\hat{F}$ considered by Bartlett (1964) and Silverman (1976, 1978). A further advantage is that for a binomial process the variance of $\hat{K}(t)$ is $2cN^{-2}$ where c depends on the shape of E but for small t is approximately $\pi t^2 A$ (the corresponding result for $\hat{F}$ is ordered $1/N$ unless toroidal edge correction is used).

Besag, in the discussion of Ripley (1977), suggested that $\hat{K}$ should be plotted on a square-root scale, i.e. the use of $\hat{L}(t) = \sqrt{(\hat{K}(t)/\pi A)}$. His motive was to linearize the plot for a Poisson process, which was also achieved by the suggestions of Cox and Kelly in that discussion, but in addition, this transformation will stabilize the variance so for a random process, $\hat{L}$ has approximate mean and variance t and $1/2\ \pi N^2$. Ripley (1977) described and exemplified tests based on $\hat{K}$. My latest results show that for a test of a binomial process (equivalently, a test of the Poisson process conditional on the number N of observations) the acceptance region of a 5% Monte-Carlo test is accurately approximated by

$$\{\sup_{r \le r_0} |L(r) - r| \leqslant 1.45/N\}$$

where r_0 is the maximum distance of interest, say 25% of the side of E if this is a square.

We have described p as the cumulative distribution function of the average distance from a point to the nearest object. This suggests that we should estimate p by an empirical distribution function. Edge effects intervene, so we proceed as follows. For m points in E we measure u_i and r_i, the distances to the nearest object and the boundary of E. Then for any process

$$\hat{p}(t) = \#(i : u_i < t, r_i < t)/\#(i : r_i < t)$$

is an unbiased estimator of $p(t)$ for t less than the maximum possible r_i. (Here # ( ) denotes the number in, and we take 0/0 to be 1.) How should we choose the m points? The obvious suggestions are a binomial process and a regular grid. Simulations suggest that it does not make much difference which is used, the random process having a slightly greater fluctuation but potentially less bias, especially on regular patterns. Certainly m should be large, at least 100 and preferably $10N$. We calculated the examples in the next section from a 30 by 30 grid although for an obviously regular pattern, I would use 1000 random points. These figures are adequate at least for up to 100 objects.

The first few inter-object distances can also be useful. Let $\nu_1 \leqslant \nu_2 \leqslant \nu_3 \ldots$ denote these distances in increasing order. Then for a binomial process Silverman and Brown (1978) show that, asymptotically for large N, $N(N-1)\pi\nu_i^2/A$ has a chi-squared distribution on $2i$ degrees of freedom. My simulation experiments show this approximation to be accurate for $N \geqslant 25$. We can use this distribution to give a one-sided test of a Binomial process against regularity, rejecting the random process for large values of ν_i. If our regular alternative is the Strauss model of non-overlapping discs, then ν_1 provides the uniformly most powerful test. For more general alternatives current work by Silverman and the author shows that ν_3 may be more suitable, since it is often considerably more powerful.

We close this section with some computational considerations. The tests based on ν_1 and ν_3 are intended to be used as quick tests for regularity with the few smallest inter-object distances located by eye. The nearest neighbour distances d_i could be found by eye, but this seems often to be inaccurate and it is better to use a computer. The machine will then have to search all inter-object distances, $O(N^2)$ computations. (In fact there are more sophisticated methods which are more efficient for large N. I find them slower for up to at least 100 objects.) This information gives us $\hat{F}$. To find $\hat{K}$ we need to compute the weights $k(x, y)$. The latter usually involve trigonometric functions and can be quite slow. However, we are usually only interested in short distances when most weights are one, so I find $\hat{K}$ takes at most twice as long as $\hat{F}$ or three times as long as CE, B_1 or B_2.

Calculation of $\hat{p}$ involves Nm operations. If m is large compared with N it may be more efficient to use the TILE algorithm of Green and Sibson (1978) to help find the nearest neighbour. With m proportional to N this method ultimately requires $O(N^{3/2})$ operations and was used here. Nevertheless, for large N we may decide to let the sampling fluctuations dominate those due to the process itself and keep m constant. Then the naive method is better; in the cases used here it would have taken up to three times longer.

Typical times on a *CDC* 6600 for $N = 100$ objects are 0.15, 0.25 and 0.75 seconds for distance methods, $\hat{K}$ and $\hat{p}$ respectively.

3.5 EXAMPLES

Our first example is not taken from geography, but it is the best I know for demonstrating the misapplication of the Clark-Evans test. It is a section of a rock with embedded magnetite objects. Davis (1973, p. 310) sets this as an exercise in the use of this test; part of the data is shown in figure 1. The uncorrected *CE* for the whole data is 2.59, whereas with Donnelly's corrections $CCE = 1.48$. It has been claimed that both statistics have a standard normal distribution. However, they give very different conclusions. The second value is consistent with simulations and gives no cause to reject a random pattern. For the portion of the data which we illustrate and use from now on, containing 38 points, $CE = 1.96$, $CCE = 0.91$ and *CCE* with toroidal distances is 1.24.

One-sided Monte-Carlo tests for regularity with B_1 and B_2 gave significance levels of 7% and 3%. In a regular pattern we would expect the nearest neighbour distances to be both larger and more equal than in a random pattern, so we reject randomness for small values of B_1 or large values of B_2. The values of $N(N-1)\pi\nu_i^2/A$ were 7.1 for ν_1 and 13.4 for ν_3, both exceeding the 95% point of the appropriate chi-squared distribution, but being less than the 99% point.

Figures 2 and 3 show plots of $\hat{L}$ and $\hat{p}$ with the envelope of 100 and 50 simulations of the Binomial process respectively. The approximate confidence band for $\hat{L}$ is also shown. Notice that $\hat{L}$ rejects randomness but $\hat{p}$ is typical of a random pattern. Let L_m denote $\sup |\hat{L}(t) - t|$ over the range of interest. Then $L_m \geqslant \nu_1$ since ν_1 is the maximum value of t for which $\hat{K}(t)$ is zero. (Remember $\hat{K}(t)$ is a weighted inter-object distance distribution function.) A *two*-sided Monte-Carlo test based on L_m had a significance level of 4%. Thus there is some evidence, not conclusive, that this map is not consistent with a 'random' process, i.e. with a binomial or Poisson process. Inspection of figure 1 suggests that there is a trend in this vertical section, more points being found at the bottom. Such heterogeneity hinders the detection of interactions.

Our second example is the set of Spanish towns analysed by Glass and Tobler (1971) and Ripley (1977). The data are 69 towns from a 'geographically homogeneous region' 40 miles square. The values of *CE* and *CCE* are 3.1 and 2.0, about 4% on a Monte-Carlo test whereas B_1 and B_2 gave significance levels of 1% and 2%. L_m calculated for the range 0–8 miles was significant in a 1% two-sided

Monte-Carlo test. Figures 4–7 show the data, the full range of $\hat{L}$, the range of interest, and $\hat{p}$. These show definite departures from randomness, there being too few pairs of towns in the range 1–4 miles apart (full scale being 40 miles, the side of the square), and $\hat{p}$ is high throughout its range, indicating regularity of spacing. The tests based on ν_1 and ν_3 are interesting, having values 6.5 and 25.0. Whereas the ν_1 value lies between the 95% and 99% points, that based on ν_3 is way beyond the 99% point. Of course any test based on ν_1 could be affected by the existence of a single pair of 'twin towns'.

Having rejected randomness I decided to try to model the effect found. I chose an inhibition process with an intertown distance of 3.5 miles, the point of the maximum deviation in $\hat{L}$. I fitted Strauss' process for various values of c, finally choosing $c = 0.5$. The plot of $\hat{L}$ for this process is shown in figure 8. Besag (1977) used an approximation procedure to estimate c as 0.7 with r equal to 4 miles. Thus our conclusion is that Spanish towns do not form a random pattern but that a mildly inhibitory mechanism is possible.

Our third example illustrates the locations of retail stores, data analysed by Rogers (1974) by fitting discrete distributions to systematic quadrat counts. The particular example shown in figure 9 is the group of 79 grocery stores in Ljubljana, Yugoslavia. The region E is a 3000 metre square, divided by Rogers into a grid of cells 50 metres square. He found that his counts were best fitted by a negative binomial distribution. He also gave values of $\bar{d}$ and $R = 0.86$. Our values of these quantities differ from his because the coordinates of the stores were taken from his figure and so are probably given to lower accuracy.

For the data given in figure 9, CCE was -3.7, significant at 1% whereas B_1 and B_2 were at about their 3% and 12% points in one-sided Monte-Carlo tests. Distance methods will always be sensitive to recording inaccuracies, particularly in tests of clustering. Figures 10, 11 and 12 show $\hat{L}$ and $\hat{p}$. These clearly reject randomness, the large values of $\hat{L}$ indicating clustering. This example shows a very similar behaviour to the Redwood data in Ripley (1977). In both, the curve of $\hat{L}$ or $\hat{K}$ rises rapidly but crosses the binomial mean curve again at 'large' distances. Note that this is not in accord with the mean curve of a Neyman-Scott cluster process which rises to a constant level (for $\hat{K}$) above that binomial mean curve. Such a process cannot be ruled out because the sampling fluctuations can be large, especially if a large number of cluster centres are allowed, but I prefer to interpret this behaviour as an indication of heterogeneity in the data. In fact figure 9 appears to represent clustering of stores in the city centre plus a fairly random distribution of stores throughout the suburbs. So the value of the analysis here or that given by Rogers is not clear.

On the other hand I believe that the analyses of regular data do have value. For even moderate numbers of points, say more than 25, the test statistics can pick out inhibition between objects which is not readily visible – to most people figure 4 looks random – and the plots of $\hat{L}$ and $\hat{p}$ can, with experience, suggest which simple mechanisms might account for the observed effect. These methods could provide a basis for the automated screening of large numbers of maps, say allocating those with significant values of L_m for later inspection. They also provide summaries by which maps can be compared and related to other variables, e.g. the regularity of large stores to the mobility of the population.

3.6 CONCLUSIONS

It is too early to draw firm conclusions from these examples. Nevertheless, the following tentative conclusions are borne out by simulation studies presently in progress. Distance methods are vunerable to the inaccurate recording of data (as is ν_1). They are quite sensitive to interaction between nearby objects, but consider only some of the information at short distances. The use of the maximum deviation of $\hat{L}$ provides a test of randomness at least as powerful as any distance method considered. The use of functional statistics such as $\hat{K}$ and $\hat{p}$ provides more insight than summary numbers, in the same way that a histogram is more illuminating than the sample mean and standard deviation. (It is of course possible to plot the object to nearest object distance distribution in the same way as $\hat{p}$; combinations of the two are being considered by Diggle. My own feeling is that there is no virtue in excluding other short distances). Since the use of a test of randomness is an indication that interaction or heterogeneity is suspected, it is surely useful that its result should indicate not only the magnitude of the effect but also its nature. Functional statistics can also be used as the basis of a method for comparing maps.

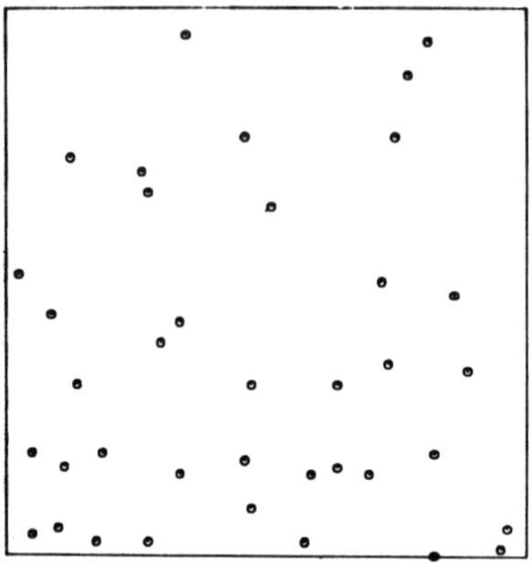

Fig. 1. Magnetite crystals in an 80 cm. square section of rock. Throughout the side of the square is the unit of distance.

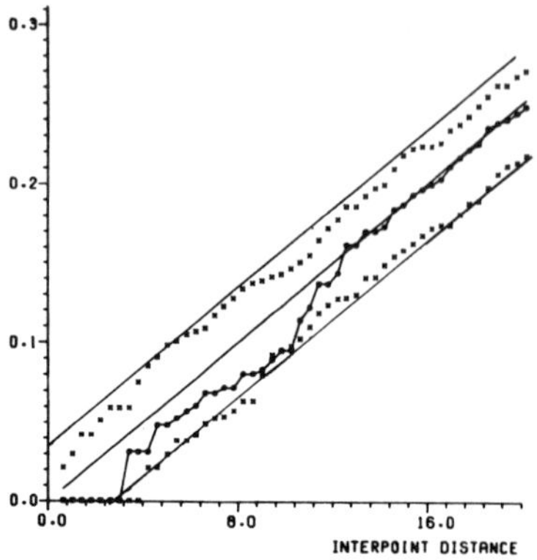

Fig. 2. Plots of $\hat{L}$ for the data of figure 1 and the envelope of 100 simulations of the Binominal process. The straight lines are the mean and 95% simultaneous confidence bands.

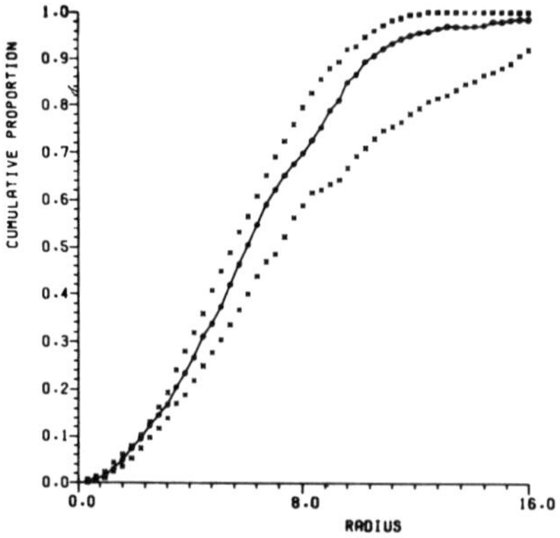

Fig. 3. $\hat{p}$ plotted for the data of figure 1 and the envelope of 50 simulations of the Binomial process.

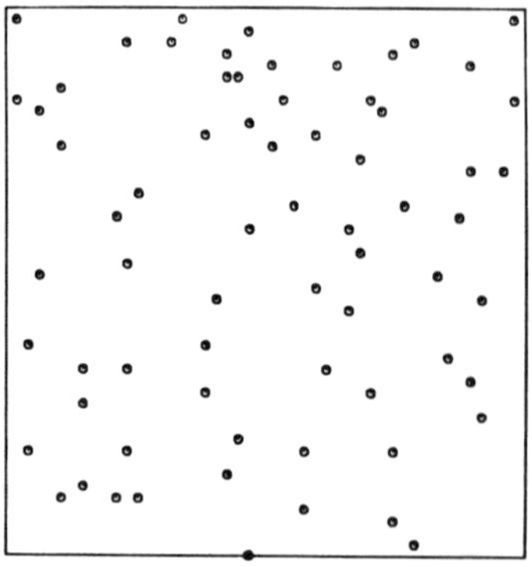

Fig. 4. The centres of 69 towns on the Spanish plain.

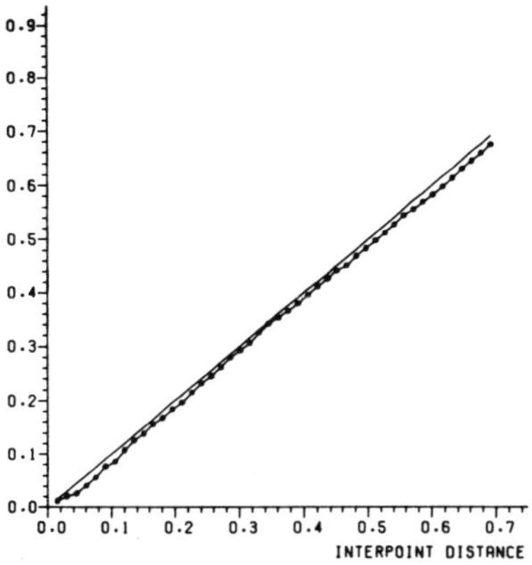

Fig. 5. Plot of $\hat{L}$ for the towns over the range of unbiasedness. The line is the mean for a Binomial process.

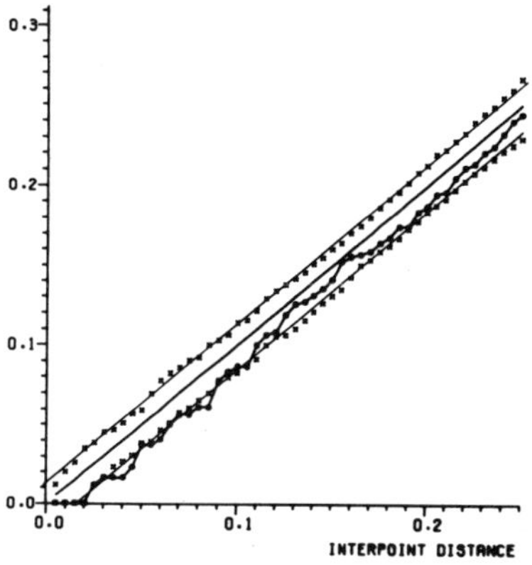

Fig. 6. $\hat{L}$ up to 10 miles for the towns and 100 simulations of the Binomial process, with the mean and 95% confidence band.

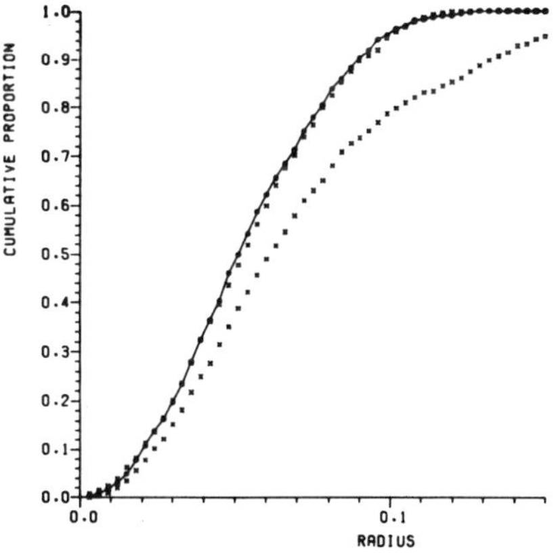

Fig. 7. $\hat{p}$ for the towns and 50 Binomial simulations.

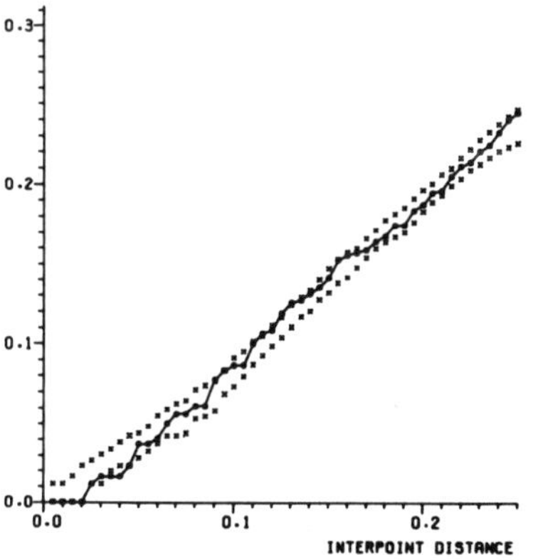

Fig. 8. $\hat{L}$ for the towns and 20 simulations of Strauss' process with $c = 0.5$, $r = 0.09$ (3.6 miles).

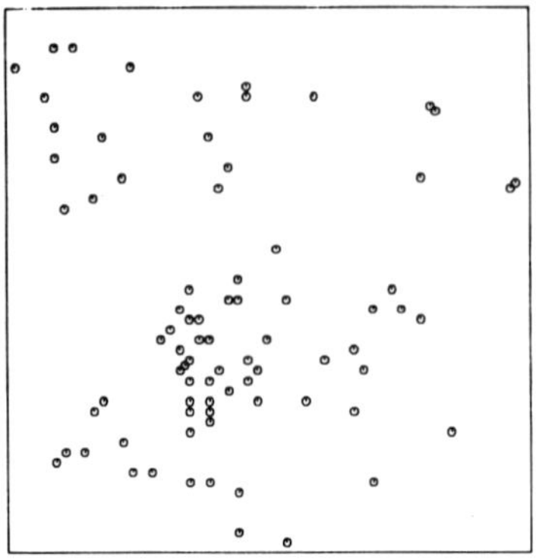

Fig. 9. 79 grocery stores in Ljubljana.

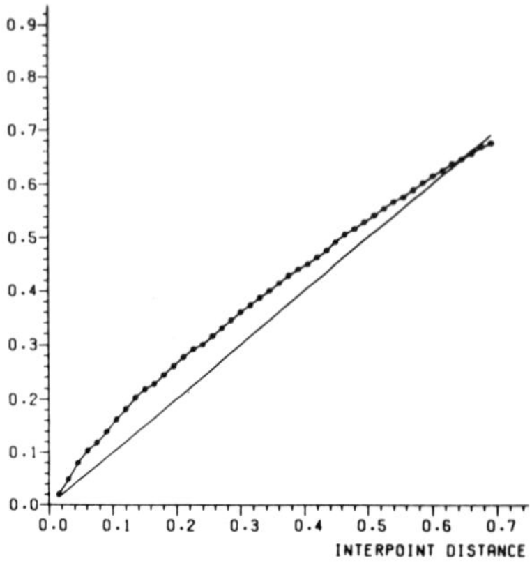

Fig. 10. $\hat{L}$ for the stores for the range of unbiasedness with the Binomial mean curve.

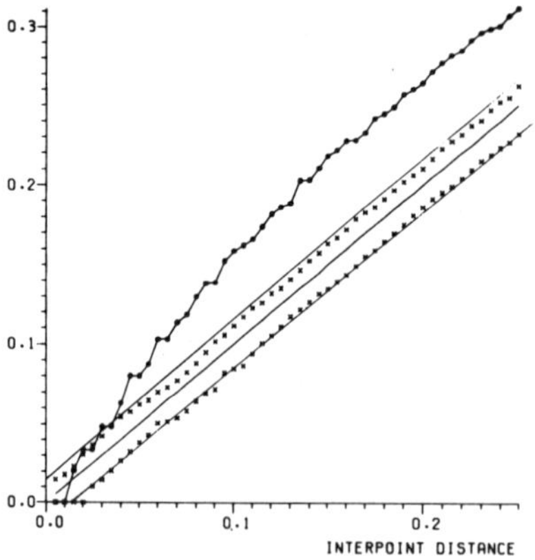

Fig. 11. $\hat{L}$ for the stores. Details as figure 2.

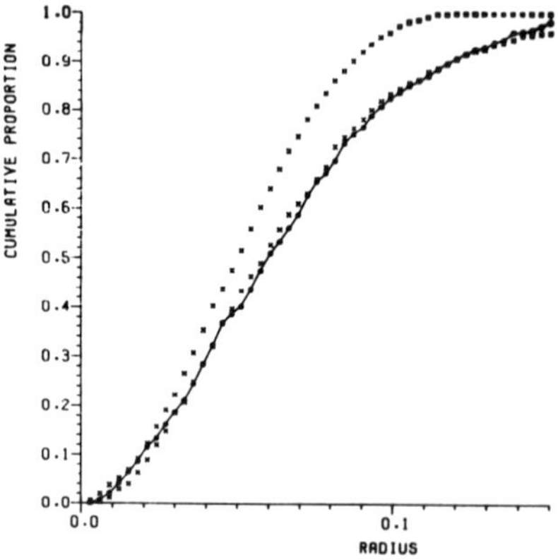

Fig. 12. $\hat{p}$ for the stores. Details as figure 3.

REFERENCES

Bartlett, M. S. (1963), 'The spectral analysis of point processes', *Journal Royal Statistical Society Series B*, 25, 264–96.

Bartlett, M. S. (1964), 'The spectral analysis of two-dimensional point processes', *Biometrika*, 51, 299–311.

Bartlett, M. S. (1975), *The statistical analysis of spatial pattern*, Chapman and Hall, London.

Besag, J. E. (1977), 'Some methods of statistical analysis for spatial data', Presented to the International Statistical Institute.

Besag, J. E. and Diggle, P. J. (1977), 'Simple Monte-Carlo test for spatial pattern', *Applied Statistics*, 26, 327-333.

Brown, D. (1975), 'A test for randomness of nest spacings', *Wildfowl*, 26, 102–3.

Clark, P. J. and Evans, F. C. (1954), 'Distance to nearest-neighbour as a measure of spatial patterns in biological populations', *Ecology*, 35, 445–53.

Cliff, A. D. and Ord, J. K. (1973), *Spatial autocorrelation*, Pion, London.

Davis, J. C. (1973), *Statistics and data analysis in geology*, Wiley, New York.

Diggle, P. J. (1976), Appendix to Hodder and Orton.

Diggle, P. J. (1977), 'The detection of random heterogeneity in plant populations', *Biometrics*, 33, 390–4.

Diggle, P. J., Besag, J. E. and Gleaves, J. T. (1976), 'Statistical analysis of spatial point patterns by means of distance methods', *Biometrics*, 32, 659–67.

Donnelly, K. (1978), 'Simulations to determine the variance and edge-effect of total nearest neighbour distance', In I. Hodder (ed.), *Simulation methods in archaeology*, Cambridge University Press, Cambridge.

Eberhardt, L. L. (1967), 'Some developments in "distance sampling" ', *Biometrics*, 23, 207–26.

Elfing, G. (1947), 'On compound binomial processes', In *Försäkrings–matematiska studier tillägnade Filip Lundeberg*, Almquist and Wicksell, Uppsala.

Errington, J. C. (1973), 'The effect of regular and random distributions on the analysis of pattern', *Journal of Ecology*, 61, 99–105.

Ford, E. D. (1976), 'The canopy of a Scots pine forest: description of a surface of complex roughness', *Agricultural Meteorology*, 17, 9–32.

Glass, L. and Tobler, W. R. (1971), 'Uniform distribution of objects in a homogeneous field: cities on a plain', *Nature*, 233 (5314), 67–8.

Goodall, D. W. (1974), 'A new method for the analysis of spatial pattern by the random pairing of quadrats', *Vegetatio*, 29, 135–46.

Green, P. J. and Sibson, R. (1978), 'Computing Dirichlet tessellations in the plane', *Computer Journal*.

Greig-Smith, P. (1952), 'The use of random and contiguous quadrats in the structure of plant communities', *Annals Botanical Society of London, New series*, 16, 293–316.

Greig-Smith, P. (1964), *Quantitative plant ecology*, 2nd ed., Butterworths, London.

Hill, M. O. (1973), 'The intensity of spatial pattern in plant communities', *Journal of Ecology*, 61, 225–35.

Hodder, I. and Orton, C. (1976), *Spatial analysis in archaeology*, Cambridge University Press, Cambridge.

Kelly, F. P. and Ripley, B. D. (1976), 'A note on Strauss' model for clustering', *Biometrika*, 63, 357–60.

Matérn, B. (1972), 'Poisson processes in the plane and related models for clumping and heterogeneity', *NATO Advanced Study Institute on Statistical Ecology*, Pennsylvania State University.

Mead, R. (1974), 'A test for spatial pattern at several scales using data from a grid of contiguous quadrats', *Biometrics*, 30, 295–307.

Newton, I., Marquiss, M., Weir, D. N. and Moss, D. (1977), 'Spacing of sparrowhawk nesting territories', *Journal of Animal Ecology*, 46, 425–41.

Pielou, E. C. (1977), *Mathematical ecology*, Wiley, New York.

Pollard, J. H. (1971), 'On distance estimation of density in randomly distributed forests', *Biometrics*, 27, 991–1002.

Ripley, B. D. (1976), 'The second-order analysis of stationary point processes', *Journal of Applied Probability*, 13, 255–66.

Ripley, B. D. (1977), 'Modelling spatial patterns (with Discussion)', *Journal Royal Statistical Society series B*, 39, 172–212.

Rogers, A. (1974), *Statistical analysis of spatial dispersion: the quadrat method*, Pion London.

Silverman, B. W. (1976), 'Limit theorems for dissociated random variables', *Advances in Applied Probability*, 8, 806–19.

Silverman, B. W. (1978), 'Distances on circles, toruses and spheres', *Journal of Applied Probability*, 15.

Strauss, D. J. (1975), 'A model for clustering', *Biometrika*, 62, 467–75.

Silverman, B. W. and Brown, T. C. (1978), 'Short distances, flat triangles and Poisson limits', *Journal of Applied Probability*, 15, 1978.

Usher, M. B. (1969), 'The relation between mean square and block size in the analysis of similar patterns', *Journal of Ecology*, 57, 505–14.

Usher, M. B. (1975), 'Analysis of pattern in real and artificial plant populations', *Journal of Ecology*, 63, 569-86.

Zahl, S. (1974), 'Application of the S-method to the analysis of spatial pattern', *Biometrics*, 30, 513–24.

4. Construction of interregional input-output tables by efficient information adding

FOLKE SNICKARS*

4.1 INTRODUCTION[1]

Much theoretical and empirical work in regional economics has been devoted to the problems of constructing and estimating intra- and interregional input-output tables (see, for example, Riefler, 1973; Paelinck and Nijkamp, 1976, for a survey). There are several reasons for this.

Firstly, input-output analysis as such is a theoretically simple and practically appealing method of treating interdependencies in an economic system. In a regional perspective, the method is increasingly useful in forecasting medium term regional growth and stagnation in view of the integration of interregional and international trade patterns.

Secondly, much effort has been devoted to the task of actually constructing interregional input-output tables, especially in view of the enormous amounts of data needed in a full interregional and intersectoral table. In fact the literature on interregional input-output models (see, for example, Riefler, 1973), contains a series of examples of assumptions made to reduce the amount of data required to make estimations tractable. There is still a lack of theoretically well underpinned explanatory models of regional trade flows (see, for example, Gordon, 1976, for an attempt at developing new theory). Although considerable progress has been made in the construction of regional input-output tables of local economies where interregional trade flows are treated as aggregates

---

*Research group for Urban and Regional Planning, Department of Mathematics, Royal Institute of Technology, S-100 44 Stockholm 70, Sweden.

1. The research was done under a grant from the Bank of Sweden's Tercentenary Fund. Lennart Ohlsson, Expert Group for Regional Development, is thanked for moral and financial assistance.

(see, for example, Morrison, 1973), somewhat less progress has been made in combining local area input-output studies with interregional trade models. The methods suggested in the current paper may be seen as contributions to this area of analysis.

In the literature a distinction is often made between survey and non-survey techniques for assessing input-output relationships. A survey method for a local area input-output study would basically amount to making use of data on the firm level for shipments of intermediaries and of consumption and investment goods (and services) to construct transaction tables. Such an approach has been used by Gerking (1976) in treating stochastic input-output models. A non-survey method would typically amount to making use of aggregate shipment data, e.g. national tables, as a starting point for making local area adjustments. Such an approach has been employed by de Kanter and Morrison (1976). The core of the current paper consists of an explicit confrontation of these two methods.

Much has been written about the problems and weaknesses of input-output analysis from an economic viewpoint (see, for example, Richardson, 1972). In this paper we will not dwell on these problems but state our assumptions without penetrating deeply into their economic logic. Instead the focus is on a case study of the statistical problem of making efficient use of different sources of information in deriving a full interregional input-output table.

In the first section of the theoretical part of the paper, we briefly outline the relations within the regional economic system. Flow balances for each sector and region are specified and their relations to aggregate production and consumption data are clarified. In the second section, (partial) survey data from Sweden on firms' sales and purchases are described and various theoretical techniques for analysing these data are presented, including regression analysis and contingency-table analysis. In the third section of the paper it is shown how a full a priori input-output matrix may be constructed by using the processed, (partial) survey data. This matrix is then confronted with existing non-survey macro data. At this stage the minimum information principle (see Snickars and Weibull, 1977), is used to derive a new input-output matrix from the a priori table, which is consistent with the macro data, by means of an efficient adding of information contained in the macro data.

In the applications part of this paper, the first section contains the results of the estimations of shipment patterns using regression analysis and contingency-table analysis. The results are briefly reviewed in an economic perspective. A second section describes the results of the full estimation procedure. In a concluding section a simple application of the interregional input-output model is demonstrated.

4.2 REGIONAL AND NATIONAL ACCOUNTS

Assume that there are m economic sectors distributed over n regions. The sectors use each others products as inputs in producing consumption and investment goods and services. They export some of their products and import other products to be used in their production processes. The production processes are operated so that a value added accrues in every sector and region. Assume that transactions are registered in real terms, measured in some fixed base-price level. The shipments made may be registered as a collection of accounts, using the following notations:

y_i^k = total input to sector i, region k

x_i^k = gross production in sector i, region k

v_i^k = value added accruing in sector i, region k

x_{ij}^{kl} = deliveries of intermediaries from sector i, region k, to sector j, region l

f_i^{kl} = shipments of finally demanded goods and services (private and public consumption, investments, inventories but not exports) from sector i, region k to region l

f_i^k = final demand deliveries from sector i, region k (no exports)

r_i^k = imports (from abroad) to sector i, region k

p_i^k = total purchases of intermediaries by sector i, region k

e_i^k = exports (to abroad) from sector i, region k

s_i^k = total sales of intermediaries from sector i, region k

y_i = total input to sector i

x_i = gross production in sector i

v_i = value added in sector i

x_{ij} = intermediary deliveries from sector i to sector j

f_i = final demand (deliveries) of products from sector i (no exports)

r_i = total imports to sector i

e_i = total exports from sector i

These notations define the basic components of interest to the current estimation framework. To sort out estimated entities we will use a tilde symbol or a hat. Thus, an estimated final demand delivery pattern will be denoted by $\tilde{f}_i^{kl}$ or $\hat{f}_i^{kl}$. We will provide an exogenously determined entity with a bar. Thus, an actual national pattern of intermediary deliveries would be denoted by $\bar{x}_{ij}$.

The regional account system has to be consistent with the way that the non-survey data to be used in the estimations are organized. Here, the data used in the Swedish medium term economic forecasts will be the main data source (see, for example, Åberg, 1971). However, the accounting system is basically a standard one encountered, for example, in Richardson 1972; Hartwick, 1971):

$$y_i^k = \sum_{j=1}^{m} \sum_{l=1}^{n} x_{ij}^{kl} + \sum_{l=1}^{n} f_i^{kl} + e_i^k \tag{2.1}$$

$$y_j^l = r_j^l + x_j^l = r_j^l + v_j^l + \sum_{i=1}^{m} \sum_{k=1}^{n} x_{ij}^{kl} \tag{2.2}$$

$$\sum_{j=1}^{m} \sum_{l=1}^{n} x_{ij}^{kl} = s_i^k \tag{2.3}$$

$$\sum_{i=1}^{m} \sum_{k=1}^{n} x_{ij}^{kl} = p_j^l \tag{2.4}$$

$$\sum_{k=1}^{n} \sum_{l=1}^{n} x_{ij}^{kl} = x_{ij} \tag{2.5}$$

$$\sum_{l=1}^{n} f_i^{kl} = f_i^k \tag{2.6}$$

$$\sum_{k=1}^{n} \sum_{l=1}^{n} f_i^{kl} = f_i \tag{2.7}$$

$$\sum_{k=1}^{n} y_i^k = y_i \tag{2.8}$$

$$\sum_{k=1}^{n} x_i^k = x_i \tag{2.9}$$

$$\sum_{l=1}^{n} v_j^l = v_j \tag{2.10}$$

$$\sum_{k=1}^{n} e_i^k = e_i \tag{2.11}$$

$$\sum_{k=1}^{n} r_i^k = r_i \tag{2.12}$$

According to (2.1) the total input to sector i, region k, may be subdivided into three basic uses: one portion of it is used up as intermediaries in the production system; another portion is delivered to final domestic uses of various kinds, while the remaining portion is exported. Imported goods and services are not discernible from domestic ones in the delivery patterns. In (2.2), the value added accruing to each sector and region is defined as the difference between the gross production and the total purchases of intermediaries. Thus, for fixed j and l the vector $(x_{1j}^{1l}, x_{2j}^{1l}, \ldots, x_{1j}^{2l}, \ldots, x_{mj}^{nl}, r_j^l)$ represents the inputs required for the current sector in that region to produce at the level x_j^l. Its direct import uses are summarized in r_j^l but also the remaining inputs may contain imports that are transferred together with other intermediary deliveries. Since imports are treated this way in the national input-output tables used in the estimations, a more elaborate earmarking of imported goods and services has not been possible here. Further, all transactions have been measured in purchaser's prices so that transport and commercial margins are referred to the receiving sector.

In this theoretical description we have disregarded the measurement problems, e.g. concerning the residuals that arise in practical assessments of the values of the accounts when starting from the production and consumption sides respectively.[2] In the application we will adjust the balances from the production side of the economic system and include the residuals in the final demand pattern and in the value added.

The primary aim of the current analysis is to estimate the intermediary shipment pattern x_{ij}^{kl}, i.e. a full interregional input-output table, from incomplete survey data. This is a formidable task in the general case, especially if the number of economic sectors and regions is large. Here we will assume that the regional and sectoral subdivision is crude enough to smooth out individual peculiarities of firms and local areas. A conceivable way of estimating the flows would then be to make use of the survey data for individual firms to build up a delivery pattern at the level of aggregation chosen. A secondary aim of the current paper is to test if the intermediary shipments, as they are measured in the survey data, fit some gravity-type regression models. If this is the case, the intermediary deliveries may be described by a rather small set of parameters, making it possible for us to calculate missing or incomplete shipment data. Such a completed delivery pattern $\tilde{x}_{ij}^{kl}$ for a certain year may not necessarily be consistent with existing national production data for that year. In terms of (2.1) –(2.12), aggregate information is certainly available from national account systems concerning $\bar{x}_{ij}$, $\bar{f}_i$, $\bar{x}_i$, $\bar{v}_j$, $\bar{e}_i$ and $\bar{r}_i$. By using industrial statistics and international trade data, $\bar{x}_i^k$ and $\bar{v}_j^l$ may be determined for industrial sectors and $\bar{e}_i^k$ and $\bar{r}_i^k$ may be gathered for all sectors. It is more difficult to find data for the production and value added in service sectors by regions in official statistics but the task is not impossible. Through studies of private and public consumption expenditures and of investments and inventory levels, information concerning $\bar{f}_i^k$ may be collected but this is typically a very time-consuming task.

Thus, assuming for the moment that there is a complete a priori estimate of the intermediary deliveries, the question arises how to proceed to make this estimate consistent with the above-mentioned official statistics concerning *actual* national (and regional) production and shipment levels. This question is relevant in view of earlier

2. See, for example, Åberg (1971) for a discussion of these problems.

experiences of constructing interregional input-output tables. In, for example, Andersson (1975), national data were broken down to a regional level without taking notice of local traits of the production system, making the estimated input-output tables strictly non-survey ones. In local input-output studies for a set of regions, detailed transaction tables are set up for each region, using as full a coverage of all shipments as possible. These survey estimates are usually not cross-checked, neither concerning the interregional shipments nor as regards their relation to existing *actual* production and input-output data on the national level. In this paper a procedure will be proposed which is capable of resolving these consistency problems, thus integrating survey and non-survey techniques.

We will not discuss this procedure in detail at this stage but merely point out the fact that there is no clear-cut distinction between actual aggregate data (provided with bars above) and estimates of such data (provided with tilde symbols above). In the current context, the production levels and value added in the service sectors form one example of data which are difficult to obtain but which have been assumed exogenous here. The regionally disaggregated final demand levels form an example of an opposite assumption made in the current study. This implies that f_i^{kl} should be included in the estimation process.

To achieve the goals set up above, we have arrived at the necessity to estimate the flow pattern of intermediaries and the flow of finally demanded goods and services, $\tilde{x}_{ij}^{kl}$, $\tilde{f}_i^{kl}$ for all sectors and regions. In the next section we will describe a Swedish set of survey data which may be used to derive these a priori estimates. This means that the analysis will concern a case study which might be seen as a less interesting approach from a general point of view. However, the data set contains trade data which are partial in very common aspects, viz. they concern only transactions involving industrial sectors. Therefore, the following discussions should be of a general interest as well.

4.3 GENERATION OF SURVEY-BASED TRANSACTION TABLES

In this section we will first describe the data collected by the Expert Group for Regional Development (ERU) in Sweden concerning industrial sales and purchases (see Ohlsson and Stighäll, 1977).

Thereafter, we will specify some regression models of regional trade patterns to be tested with these data. Finally, we will present some multiplicative models of shipment patterns to be tested by contingency table analytic methods (see, for example, Kullback, 1959; Haberman, 1976).

4.3.1 *Swedish shipment data*

In the survey which was performed in 1973, data were collected about sales and purchase patterns, energy use, etc. for all industrial sectors. The *total* sales to domestic clients as well as total export figures were collected. Thus, no subdivision was made between sales of intermediary goods and services and sales of consumption and investment goods. Furthermore, the purchases of *intermediaries* made by the industrial sectors from all domestic clients were registered together with their total imports.

Regarding the domestic transactions, a complete coverage of all purchases of intermediaries would at the same time serve as a registration of all sales of intermediaries since accounts are kept of flows both at the sender and at the receiver. Then the total flows of final demand goods and services could easily be derived.

However, in the delivery survey only firms with more than 50 employees were included. Furthermore, these firms were asked to state the purchases from their 15 largest clients and the sales to their 10 largest clients in addition to the total sales and purchases of intermediaries. This implies that the sales data contain some 75 per cent of the production values in most counties in Sweden and also on the national level. For the purchases, the corresponding figures, which are valid for industry as a whole, are some 65 per cent. The span in coverage is 60–96 per cent for sales and 52–84 per cent for purchases. This means in summary, that the data are not random samples and that the coverage varies considerably over sectors and regions, at least for the finest regional and sectoral subdivision available.

The contents of the data sets are summarized in Figure 1, where m_I and m_S are the number of industry and service sectors, respectively, and thus $m_I + m_S = m$. It should be observed that the firms have stated their total sales and purchases together with exports and imports independently of the more detailed data on intermediaries. The latter data concerns only the larger shipments made. This means that local, mostly small deliveries are to a large extent absent from the data. The geographical implications of these facts

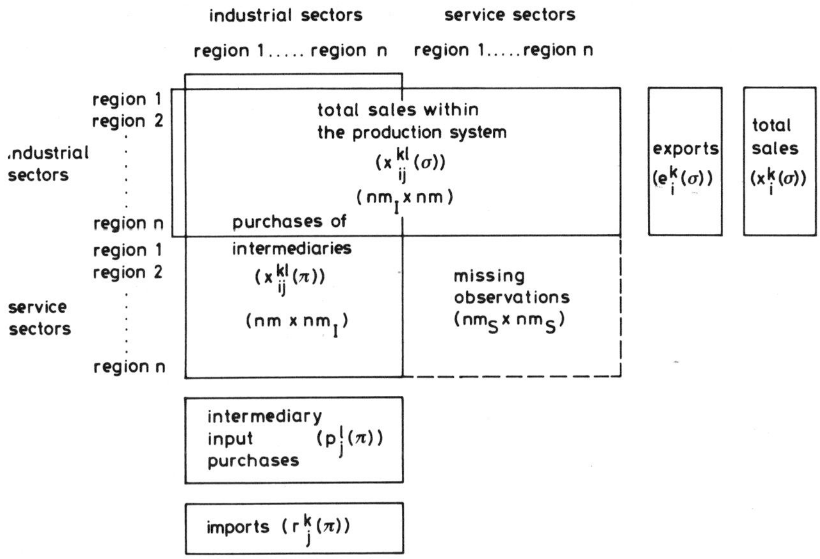

Fig. 1. Principle structure of the data sets. The notations (σ) and (π) stand for total sales and intermediary sales data, respectively.

should be kept in mind when analyzing the data. Another important fact to be kept in mind is that the delivery data do not reflect the (geographical) distribution of the actual physical deliveries but rather of the payments for these deliveries. This should tend to over-estimate the importance of metropolitan areas in the delivery network because of the location of many head offices of large firms with distributed organizations to those areas.

As may be gathered from Figure 1, no data are available on the transactions between service sectors. The aim of the regression and contingency table analyses is to use the survey data available to obtain an estimate of the full delivery patterns of intermediaries and final demand products. The regression analyses will be made primarily to point at regularities in the transaction patterns in a qualitative way, whereas the contingency-table approach is intended to provide the basis for the final estimation step. In view of the existing non-survey data, which are rather detailed concerning intersectoral linkages, the subsequent analysis will focus on the *spatial* characteristics of the trade flows. Do geographical distances (transport costs) play an important role in determining shipment patterns? If so, are there significant differences in the distance sensitivity of shipments of products pertaining to different economic sectors? Is it true that there is a typical interregional delivery pattern for each sending sector and region, regardless of which sector receives these shipments?

4.3.2 *Some gravity type models of trade flows*

Studies of trade flow patterns usually employ gravity type models as descriptive tools, for instance Leontief and Strout (1966), Black (1972) and O'Sullivan and Ralston (1974). However, there also exist trade flow models based on technology assumptions of economic theory, (see, for example, Stilwell and Boatright, 1971). For a more elaborate model of transaction tables from the point of view of econometric theory, see Streitberg (this volume). Most empirical approaches amount to a multiplicative specification of the effects of the sizes of the sending and receiving sectors and their geographical locations on the shipment patterns of primary and intermediary products. In more recent approaches, the estimations are made with various linear constraints on the elements of the transaction tables (see, for example, Openshaw, this volume, for several examples using commuting data and Bartels, this volume, for a critical review of these methods from a statistical point of view).

Here we will specify two gravity-type models and estimate them free of restrictions. The main reason for this simplification is that the analysis is intended to be exploratory. Let d^{kl} denote the distance between regions k and l, serving as a reflection of transport costs, and let x_{ij}^{kl} be the shipment pattern to be modelled. The following shipment model introduces general multiplicative relations between deliveries, production levels and transport costs:

$$x_{ij}^{kl} = e^{\rho_i} \cdot (x_i^k)^{\alpha_i} \cdot (x_j^l)^{\beta_i} \cdot e^{-\gamma_i d^{kl}} \cdot e^{\epsilon_{ij}^{kl}} \qquad (3.1)$$

According to (3.1) there is a set of elasticity parameters $(\alpha_i, \beta_i, \gamma_i)$, varying by sending sector only. Equation (3.1) may be regarded as a 'production function' for intermediary inputs to sector j in region l from sector i in region k. Then, $\gamma_i d^{kl}$ measures the percentage decrease in the shipment level if the (road) distance between regions k and l is increased by one percent. In the same fashion $(\alpha_i + \beta_i)$ are production scale parameters, expressing the elasticity of the shipment level to a change in each of the two production levels involved. Lastly, α_i and β_i are the corresponding elasticities with respect to changes in the production levels of the input and output sectors, respectively. For a fixed receiving sector and region (j, l) the vector

$$\Big\{ e^{\rho_1} (x_1^1)^{\alpha_1} (x_j^l)^{\beta_1} e^{-\gamma_1 d^{1l}}, e^{\rho_1} (x_1^2)^{\alpha_1} (x_j^l)^{\beta_1} e^{-\gamma_1 d^{21}}, \ldots$$

$$\ldots, e^{\rho_2} (x_2^1)^{\alpha_2} (x_j^l)^{\beta_2} e^{-\gamma_2 d^{1l}}, \ldots,$$

$$e^{\rho_m} (x_m^n)^{\alpha_m} (x_j^l)^{\beta_m} e^{-\gamma_m d^{nl}} \Big\} \tag{3.2}$$

summarizes the pattern of intermediary inputs. This should be compared to the linear model of production (see, for example Nikaido, 1968) where the input vector would be

$$\Big\{ a_{1j}^{1l} x_j^l, a_{1j}^{2l} x_j^l, \ldots, a_{2j}^{1l} x_j^l, \ldots, a_{mj}^{nl} x_j^l \Big\} \tag{3.3}$$

(3.2) is seen to be a much more elaborate model which may reveal both decreasing and increasing returns to scale in the production of transaction flows, depending on the value of $(\alpha_i + \beta_i)$. In the linear model (3.3) constant returns prevail by definition of the constant interregional and intersectoral input-output coefficients a_{ij}^{kl}. It is of considerable interest to judge which case holds here.

In the regression equation (3.1) a stochastic error factor has been specified. The assumptions to be made about its distribution depend on the properties of the data sets available for the estimations. As noted by e.g. Bartels (this volume), an assumption of ϵ_{ij}^{kl} being normally distributed will lead to biased forecasts, so the issue is an important one. As our aim in this section is not primarily to make projections, we will assume normality to hold, thus justifying application of least-squares estimation.

In the current case, production data by sector and region are only available for industrial sectors. From Figure 1 we see that two different data sets are available for the estimations. The patterns of total sales for different *industrial* sectors may be studied by employing the data set $\{x_{ij}^{kl}(\sigma), x_i^k(\sigma), x_{:j}^{\,:l}(\sigma)\}$ where $x_{:j}^{\,:l}(\sigma)$ represents the sum of $x_{ij}^{kl}(\sigma)$ over sending sectors and regions and thus deliveries made to all sectors are included. The patterns of sales of intermediary products from *all* sectors to the industrial sectors may be studied by making use of the data set $\{x_{ij}^{kl}(\pi), x_{i:}^{k:}(\pi),$

$x_j^l(\sigma)\}$, where $x_{i\cdot}^{k\cdot}(\pi)$ stands for the sum of $x_{ij}^{kl}(\pi)$ over all industrial sectors and regions. In the latter regressions there are no restrictions imposed on the estimated flows, although they should actually sum to $x_{i\cdot}^{k\cdot}(\pi)$ by the definition of that entity. In view of the other weaknesses of the data set, it has not been judged mandatory to include these restrictions.

In (3.1) no explicit notice is taken of specific intersectoral relationships. We know from national input-output data that such relationships are very marked, e.g., some intersectoral flows are practically zero. Such effects would be grossly misjudged by the specification (2.1). To specify a regression model that lets the level constant vary with the intersectoral constellation (i, j), let the dummy variable y_j identify sector j so that $y_j = 1$ if sector j stands as the receiver of a certain shipment and $y_j = 0$ otherwise. Then we may rewrite (3.1) as follows:

$$x_{ij}^{kl} = e^{\sum_{j=1}^{m} \rho_{ij} y_j} \cdot (x_i^k)^{\alpha_i} \cdot (x_j^l)^{\beta_i} \cdot e^{-\gamma_i d^{kl}} \cdot e^{\epsilon_{ij}^{kl}} \qquad (3.4)$$

By rewriting them in logarithmic form, both (3.1) and (3.4) may be estimated by standard regression techniques. Some results of such estimations, using (3.4), will be briefly presented below. There, the focus will be on a comparison between the distance dependency of the shipment patterns for different sectors and also on the differences between the transport cost sensitivity of intermediary products delivered by a sector and that of shipments of finally demanded goods from the same sector. By comparing $\tilde{\gamma}_i(\sigma)$ with $\tilde{\gamma}_i(\pi)$ conclusions might be drawn that help in constructing a full intersectoral and interregional shipment matrix.

4.3.3 *Interaction effects in contingency tables*

In the literature (see, for example, Kullback, 1959; Haberman, 1976), a contingency table is essentially defined as a sample of cell count frequencies from a population which may fall into various categories. In testing log-linear models applied to these tables, it is usually assumed that the sample is drawn from one or several independent multinomial distributions. There are numerous applications of contingency-table analysis in the social sciences (see Haberman, 1976, for a survey). Among applications dealing with transaction tables, studies by Goodman (1963), using international trade data, may be mentioned.

Here, it is argued that the approach is useful in detecting interaction patterns between economic sectors and regions without demanding that these interactions should exhibit certain assumed functional forms. One example of this is that interregional shipment effects can be discerned without pre-specifying that these interactions should a priori be simple functions of geographical distance. Such a statement about distance effects is especially pertinent if there is a coarse regional subdivision and the distance measures are crude.

In the former section we specified two regression models, which were intended to summarize the patterns of total sales and sales of intermediaries for each sending sector. In this section we will specify a similar model of log-linear type, viz.

$$x_{ij}^{kl} = A_{ij}^{l} \cdot B_{i}^{kl} \tag{3.5}$$

The model (3.5), in a way, contains the two models (3.1) and (3.4) as special cases. According to (3.5) the shipments of intermediaries can be decomposed into the product of two independent interaction effects. The factor A_{ij}^{l} might be termed the technology effect since it makes up for an *input* technology which is typical for each receiving sector and region. The factor B_{i}^{kl} might be termed the interregional transport effect since it reflects the assumption that there is a typical interregional delivery pattern for the *output* of each sector. This interregional effect is not tied to distance or transport costs alone but may as well contain interaction patterns that are due to the ownership structure of firms or the historically established delivery pattern. Basic points of interest are, therefore, to compare the regionally differentiated input-output technology effect A_{ij}^{l} with the national intersectoral linkages $\overline{x}_{ij}$ and to compare the regional "metric" inherent in B_{i}^{kl} with the distance measure d^{kl}. An equivalent specification of model (3.5) is

$$x_{ij}^{kl} = x_{i}^{k} \, x_{j}^{l} \, a_{ij}^{l} \, b_{i}^{kl} \tag{3.6}$$

where the magnitudes of the sending and the receiving sectors enter explicitly.

Let $N_l(\sigma)$ and $N_l(\pi)$ be subsets of the indices (j, k, l) with $\overline{x}_{ij} > 0$, i.e. such that there may actually be non-zero deliveries present in the samples, and such that deliveries from sector i to sector j are actually registered in the data (recall the limitations concerning service sectors mentioned in section 3.1).

For an index set N_i of the type stated above, assume that q_{ij}^{kl} for a given $i = 1, \ldots, m$ is the probability that a transaction will take place between region k and sector j in region l. Given a sample size X the model (3.5) may also be expressed in terms of the probabilities $q_{ij}^{kl} > 0$ as

$$x_{ij}^{kl} = X \cdot q_{ij}^{kl} = A_{ij}^{l} B_{i}^{kl} \tag{3.7}$$

for both the σ – and the π – versions of the transaction samples.

A crucial assumption here is that the multinomial distribution assumptions are really valid for transaction flow data. The sample sizes are measured in monetary terms here (in units of 1000 Skr) so that the cell counts actually consist of the number of monetary units delivered in the various transactions. However, in spite of these problems, contingency models have been proposed to analyse shipment data (see, for example, Savage and Deutsch, 1960; Goodman, 1963).

In the current application, it would have been preferable to analyse each of the samples $\{x_{ij}^{kl}(\sigma)\}$ and $\{x_{ij}^{kl}(\pi)\}$ in a single step. Because of the fact that we know a priori that some flows should be identically zero, the assumptions that $q_{ij}^{kl}(\sigma) > 0$ and $q_{ij}^{kl}(\pi) > 0$ would be violated in a full scale analysis. By constraining the analysis to deal with each sending sector separately, this problem is circumvented and the zero elements are left outside the analysis. Another advantage with this approach is that the analysis becomes comparable to the regression model estimations of section 4.3.2.

Under the assumptions stated above it may be shown that maximum likelihood estimation yields unique estimates of the elements in the contingency tables (see Haberman, 1976). Solution of the likelihood equations is generally accomplished by so called iterative proportionate fitting or by Newton-Raphson techniques. One of the main advantages of log-linear hierarchical models of the type (3.5) is the simplicity of estimation. Using iterative proportionate fitting (see, for example, Bacharach, 1970), large tables may be analyzed at moderate costs.

The results of these estimations will be two sets of parameter estimates, one for each of the data sets denoted by σ and π, i.e.

$\{\tilde{A}^{l}_{ij}(\sigma), \tilde{B}^{kl}_{i}(\sigma)\}$ and $\{\tilde{A}^{l}_{ij}(\pi), \tilde{B}^{kl}_{i}(\pi)\}$. With the help of these estimated parameters, the missing elements of the table of intermediary deliveries may be filled in. One possible approach to generate such an a priori estimate of the shipment patterns, which may not fulfill all exogenous restrictions, will be given in the next section together with a method for achieving overall consistency.

4.3.4 *Application of the minimum-information principle*

The aim of this section is to account for the procedure adopted in tying the partial approaches of the former sections together. The first task is to construct a full intersectoral and interregional shipment matrix from the model estimations presented above. Here, these a priori estimates will all be based on the contingency table approach. In fact, the estimates derived by using the intermediary shipment sample will be used for all deliveries of intermediary products, i.e.

$$\tilde{x}^{kl}_{ij} = \tilde{\tilde{A}}^{l}_{ij}(\pi) \cdot \tilde{B}^{kl}_{i}(\pi) \tag{3.8}$$

where

$$\tilde{\tilde{A}}^{l}_{ij}(\pi) = \begin{cases} \tilde{A}^{l}_{ij} & \text{if } (j, l) \in N_i(\pi) \\ 1 & \text{if } (j, l) \notin N_i(\pi) \text{ but } \bar{x}_{ij} > 0 \\ 0 & \text{if } \bar{x}_{ij} = 0 \end{cases} \tag{3.9}$$

(3.8) reveals that the current interregional input-output model is of the type proposed by Chenery (1956) and Moses (1955) (see also Riefler, 1973). It is somewhat more intricate to derive an estimate of the domestic deliveries of final demand products. Observe that the shipments of exported products within the country are not included here. Since these shipments are neither present in the sample $\{x^{kl}_{ij}(\sigma)\}$ nor $\{x^{kl}_{ij}(\pi)\}$ and they have a distinct direction towards export harbours as opposed to other final demand shipments, they have been excluded and treated as exogenous data.

In a possible attempt to derive a priori estimates $\tilde{f}^{kl}_{i}$ of domestic final demand shipments, we have assumed that the exogenous data summarized in section 4.2 also contain a preliminary judgement of final demand levels specified by sectors and regions of the type $\upsilon^k \bar{f}_i$ where υ^k may, for instance, be given by

$$\upsilon^k = \sum_{i=1}^{m} \bar{v}_i^k \Big/ \sum_{i=1}^{m} \sum_{k=1}^{n} \bar{v}_i^k .$$

The assumption is that the interregional distribution of the final demand follows that of the value added accruing in different regions. Using this crude assumption the estimated shipments $\tilde{x}_{ij}^{kl}(\sigma)$ may be partitioned into intermediary shipments $\tilde{x}_{ij}^{kl}(\pi)$ and domestic final demand shipments $\tilde{f}_i^{kl}$. Normalizing the shipment levels so that the preliminary final demand levels arise on aggregation, we have

$$\tilde{f}_i^{kl} = \frac{(\bar{x}_i^k - \bar{e}_i^k)}{g_i^k(\sigma)} \sum_{j=1}^{m} \tilde{\bar{A}}_{ij}^{l}(\sigma)\, \tilde{\bar{B}}_i^{kl}(\sigma) - \frac{(\bar{x}_i^k - \bar{e}_i^k - \upsilon^k \bar{f}_i)}{g_i^k(\pi)} \sum_{j=1}^{m} \tilde{\bar{A}}_{ij}^{l}(\pi)\, \tilde{\bar{B}}_i^{kl}(\pi) \tag{3.10}$$

where

$$g_i^k(\cdot) = \sum_{j=1}^{m} \sum_{l=1}^{n} \tilde{\bar{A}}_{ij}^{l}(\cdot)\, \tilde{\bar{B}}_i^{kl}(\cdot) \tag{3.11}$$

and $\tilde{\bar{A}}_{ij}^{l}(\sigma)$ is defined in the same way as $\tilde{\bar{A}}_{ij}^{l}(\pi)$ (see equation (3.9), and $\tilde{\bar{B}}_i^{kl}(\sigma) = \tilde{\bar{B}}_i^{kl}(\pi)$ for i-values pertaining to service sectors and the primary sector). By summing (3.10) over l the (positive) final demand estimates $\upsilon^k \bar{f}_i$ arise. However, it cannot be excluded that some additional restrictions may have to be introduced to guarantee that the crude formula (3.10) will yield positive estimates throughout.

The final step in the estimation procedure consists in confronting the estimates $(\tilde{x}_{ij}^{kl}, \tilde{f}_i^{kl})$ with the exogenously given data by using information-theoretic methods. Following Jaynes (1957), Kullback (1959), Hobson (1971) and Snickars and Weibull (1977), we introduce the following measure of the extra information content of the matrix (x_{ij}^{kl}, f_i^{kl}) fulfilling (2.1) – (2.12) in relation to that of $(\tilde{x}_{ij}^{kl}, \tilde{f}_i^{kl})$:

$$I((x, f), (\tilde{x}, \tilde{f})) = \sum_{i=1}^{m} \sum_{j=1}^{m} \sum_{k=1}^{n} \sum_{l=1}^{n} x_{ij}^{kl} \log\left\{\frac{x_{ij}^{kl}}{\tilde{x}_{ij}^{kl}}\right\} + \sum_{i=1}^{m} \sum_{k=1}^{n} \sum_{l=1}^{n} f_i^{kl} \log\left\{\frac{f_i^{kl}}{\tilde{f}_i^{kl}}\right\} \tag{3.12}$$

In information theory, Hobson (1971), has derived (3.12) as a unique measure of the information content in an additional message (x, f) given the message $(\tilde{x}, \tilde{f})$ by extending the axioms used by Shannon and Weaver (1949) in defining the entropy concept axiomatically. In statistics Kullback (1959), among others, has used expressions like (3.12) as the logarithm of the likelihood ratio in a random sample situation. In dealing with extensions of entropy models, Snickars and Weibull (1977) have shown that (3.12) can be given a probabilistic underpinning even if random samples are not at hand. By seeking the minimum of $I((x, f), (\tilde{x}, \tilde{f}))$, a distribution (x, f) is sought that lies as 'close' to $(\tilde{x}, \tilde{f})$ as possible in an information-theoretic sense (see also Hårsman and Marksjö, 1977). Here we will use this technique to derive a 'conservative' estimate of a full interregional input-output table which is consistent with existing non-survey data.

This may be done by solving the following mathematical programme:

$$\min I((x, f), (\tilde{x}, \tilde{f}))$$

with respect to

$$\{x_{ij}^{kl}\}, \{f_i^{kl}\}$$

subject to

$$\sum_{j=1}^{m} \sum_{l=1}^{n} x_{ij}^{kl} + \sum_{l=1}^{n} f_i^{kl} = \bar{x}_i^k + \bar{r}_i^k - \bar{e}_i^k \tag{3.13}$$

$$\sum_{i=1}^{m} \sum_{k=1}^{n} x_{ij}^{kl} = \bar{x}_j^l - \bar{v}_j^l \tag{3.14}$$

$$\sum_{k=1}^{n} \sum_{l=1}^{n} x_{ij}^{kl} = \bar{x}_{ij} \tag{3.15}$$

$$\sum_{k=1}^{n} \sum_{l=1}^{n} f_i^{kl} = \bar{f}_i \tag{3.16}$$

$$x_{ij}^{kl} \geqslant 0, \; f_i^{kl} \geqslant 0 \tag{3.17}$$

It may be shown that there is always a unique solution to the above minimisation problem, which may be expressed analytically in terms of a set of monotone transformations of the Lagrange multipliers pertaining to (3.13) – (3.16). Since all constraints are equalities, it may also be shown that the individual Lagrange parameter values are not unique. The solution is given by

$$\hat{x}_{ij}^{kl} = \lambda_i^k \cdot \varkappa_{ij} \cdot \tilde{x}_{ij}^{kl} \tag{3.18}$$

$$\hat{f}_i^{kl} = \lambda_i^k \cdot \varphi_i \cdot \tilde{f}_i^{kl} \tag{3.19}$$

where λ_i^k, μ_j^l, $\varkappa_{ij}$ and φ_i pertain to restrictions (3.13), (3.14), (3.15) and (3.16) respectively. The solution (3.18) – (3.19) may be obtained numerically by iterative proportionate fitting as in the former section (see Bacharach, 1970; Ireland and Scheuren, 1974), wo provide efficient methods for doing this).

Observe that if $\tilde{x}_{ij}^{kl} = 0$ or $\tilde{f}_i^{kl} = 0$ the corresponding shipment estimate will always be zero. Furthermore, it may be observed that it is futile to put down work in modifying the a priori forecast by factors of the *same* type provided by the transformed Lagrange multipliers. The effects of such modifications will be completely wiped out in the final estimates. Thus, it does not mattter, for example, whether $\tilde{x}_{ij}^{kl}$ or $\xi_i^k \cdot \tilde{x}_{ij}^{kl}$ for some factor ξ_i^k are used as a priori forecasts. In practical work with the method, this is often a very useful observation.

The above estimation procedure has a practical appeal, being relatively easy to perform by computer. In this context, we will not analyse its economic content but rather look upon it as an information-theoretical approach to a practical statistical estimation problem. It forces the analyst to specify his pieces of information and provides him with estimates that are 'conservative' with regard to that information.

4.4 RESULTS OF THE STATISTICAL ESTIMATIONS

In this section we will present the main results of estimations of the regression model (3.4) and of the log-linear model (3.5). We will focus on the spatial implications of the results and on the estimations dealing with scales of intermediaries.

Although the original data collected in the 1973 delivery survey (see Ohlsson and Stighäll, 1977), were assembled at the firm level, it has only been possible in this context to analyze these data at a considerably more aggregate level. Thus the firms have been grouped into 19 sectors of which 13 are industrial ones, 5 are service sectors and the remaining one is the primary sector. The choice of sectors has been made so as to match the subdivision used in the national economic planning in Sweden (see Table 1, sec. 4.4.1, for an identification of the sectors). A regional subdivision into 8 regions has been employed, consisting of groupings of geographically adjacent counties. These county groups are used in governmental regional planning in Sweden. They have been derived to a certain extent on transportation cost grounds but also so as to yield regions of comparable population size (see Figure 2). However, no precise criteria regarding regional economic structure underpin this regional subdivision.

4.4.1 *Regression results*

The results of the regressions suffer from the basic fact that individual observations have not been available in the current study, although such data are present in the background material. On top of this weakness, there is a problem of bias in the estimations relative to the true values of the population at study due to the structure of the samples.

Trying to estimate the shipment patterns for groups of *sending* sectors, a bias is introduced in the current sample relative to a

Number	Name
1	Stockholm county
2	East central Sweden
3	Southeast Sweden
4	Southern Sweden
5	Western Sweden
6	West-central Sweden
7	Lower Norrland
8	Upper Norrland

Fig. 2. Regional subdivision of Sweden

random one, by the facts that only firms with more than 50 employees were members of the sample and that these firms were asked to state their sales to the 10 largest clients (in production value) and their purchases (of intermediaries) from the 15 largest clients. It might be guessed that the current estimates would tend to overestimate economies of scale and underestimate the distance sensitivity. This guess is based on the assumption that small firms generally purchase and sell more locally than larger ones and that they do not exhibit very marked scale economies.

Knowing this, the regressions dealing with sales of intermediaries by industrial sectors (data set π) may be analyzed in a rather straightforward way. Concerning the total sales regressions (data set σ), the results are further obscured by the facts that it is impossible to discern in detail how large the sales of intermediaries are in relation to the sales of investment and consumption goods. As regards investment products, we know that they are sold only by a few sectors, primarily by metal manufacturing firms and construction companies. Concerning consumption goods, we know that only deliveries made to other sectors in the production system are registered. Since public sectors and individual households are excluded, the amounts involved need not be too large.

In the presentation of the results of the estimations[3] we will focus on the distance 'elasticities' (γ_i) and on the production scale elasticities ($\alpha_i + \beta_i$). The regressions based on (3.4) will primarily be analyzed.

The main results of the estimations using data set (π), i.e. the deliveries of intermediary products, are summarized in Table 1.

As may be concluded from Table 1, R^2-values vary between 0.30 and 0.77 for the regression model (3.4). In some cases, however, the sample sizes are too small to get reliable estimates. It might be noted that the dummy variables y_j of (3.4) add a considerable explanatory power if compared with model (3.1). If the latter model specification is used in the estimations, the regressions yield R^2-values in the range 0.03 – 0.42. It may be concluded that the intersectoral technology effect is quite important in the shipment patterns. Model (3.1) gives a rather poor description of delivery patterns in general.

Concerning distance 'elasticities', we see from Table 1 that 12 out of 19 sectors yield significantly positive values at the 95 percent level. Only two sectors have (unsignificantly) negative values. According to the results there are considerable differences between the distance dependency of deliveries of intermediaries made from different economic sectors. The ones that produce agriculture and food products as intermediary inputs to other sectors reveal the strongest distance dependency. Thereafter, a subgroup is formed by the sectors that deliver paper products in a wide sense. An unexpected result is that the service sectors tend to have lower distance elasticities in their service deliveries than some of the industrial sectors. Apparently, industrial sectors use service inputs not only from their own county group but also from a subset of the others. It may be remarked that the above result may also be interpreted in the opposite way, viz. as an indication that the industrial sectors do not deliver a homogeneous good (or goods) as output but maybe also a certain amount of services (e.g. in firms with several production facilities).

As noted in section 4.3.1 the shipment formulas (3.1) and (3.4) may also be looked upon as 'production functions' for intermediary shipments. The sum ($\alpha_i + \beta_i$) contains information about how the shipments depend on the production scale of the two sectors in the

---

[3] Anita Sjöstedt, Central Bureau of Statistics, is thanked for computational assistance.

Table 1. Regression results using data set (π). (Rank order of sectors according to distance and production scale elasticities).

Sector (i)	N_i	R_i^2	$\gamma_i \cdot 10^{-3}$	Rank order	$(\alpha_i + \beta_i)$	Rank order
(1) Agriculture and forestry	70	0.42	5.747*	1	0.695**	6
(2) Mining and quarrying	82	0.68	0.633	13	0.895**	10
(3) Food manufacturing	99	0.65	3.421*	2	0.952*	14
(4) Textile and clothing	144	0.68	0.724*	12	1.000**	15
(5) Wood and paper products	344	0.48	1.989*	4	0.450**	2
(6) Printing and publishing	96	0.66	2.198*	3	0.474**	3
(7) Rubber products	116	0.61	0.122	17	1.901**	11
(8) Chemical industry	374	0.49	1.046*	10	1.293**	18
(9) Petroleum and coal products	120	0.59	0.594	15	1.146**	16
(10) Non-metallic mineral products	141	0.47	0.627*	14	0.914**	12
(11) Basic metal industries	203	0.44	0.462*	16	1.487**	19
(12) Machinery and equipment	348	0.54	1.517*	6	0.917**	13
(13) Shipbuilding	21	0.72	−1.259	19	0.571*	4
(14) Other manufacturing	118	0.65	−0.466	18	0.259**	1
(15) Electricity, gas, water, etc	53	0.77	0.786	11	0.701**	7
(16) Construction	79	0.47	1.179*	8	0.821**	9
(17) Wholesale and retail trade	358	0.30	1.642*	5	1.261**	17
(18) Transport and storage	58	0.55	1.449	7	0.707**	8
(19) Other private services	91	0.52	1.107*	9	0.644**	5

Notation: N_i = number of observations
R_i^2 = R-square for the model
* = significance at the 95 percent level
** = both α_i and β_i significant

two regions involved. In the interpretation of these parameter estimates, it should be observed that total production values have only partly been used in the estimations, due to the structure of the data sets. Since $(\alpha_i + \beta_i)$ relate to percentage marginal changes the effect of this should not be too marked, however. From Table 1 it is seen that all α– and β–values but two are significantly different from zero at the 95 percent level. For 15 out of 19 sectors the marginal percentage increase in intermediary shipments at a production scale increase of one percent is below or equal to one. For three sectors, two of which produce paper products in a wide sense, it is below 0.5. At the other end of the scale, shipments of intermediary products from the basic metal industries increase some 1.5 percent at percentage production scale increases.

Although these results are somewhat difficult to interpret, they indicate that for most sectors the assumption of constant input-output coefficients is not too weak, although for some sectors it seems to be severely violated. The results of the regression estimations using data set σ, comprising total sales patterns for industrial sectors, are summarized in Table 2.

Table 2. Regression results using data set (σ). (Rank order of sectors according to distance and production scale elasticities).

Sector (i)	N_i	R_i^2	$\gamma_i \cdot 10^{-3}$	Rank order	$(\alpha_i + \beta_i)$	Rank order
(2) Mining and quarrying	77	0.71	0.511	9	0.667**	2
(3) Food manufacturing	125	0.49	3.230*	1	1.001**	7
(4) Textile and clothing	214	0.56	1.438*	5	1.153**	10
(5) Wood and paper products	421	0.42	1.248*	7	0.841**	3
(6) Printing and publishing	180	0.65	1.380*	6	0.895**	4
(7) Rubber products	91	0.71	0.143	10	1.295**	11
(8) Chemical industry	347	0.43	0.709*	8	1.041**	8
(9) Petroleum and coal products	16	0.97	2.152	2	1.356	12
(10) Non-metallic mineral products	204	0.54	1.471*	3	0.600**	1
(11) Basic metal industries	157	0.44	-0.210	12	1.435**	13
(12) Machinery and equipment	624	0.45	1.448*	4	0.995**	6
(13) Shipbuilding	78	0.81	–0.427	13	0.930**	5
(14) Other manufacturing	88	0.81	0.065	11	1.080**	9

Notation: N_i = number of observations
R_i^2 = R–square for the model
* = significance on the 95 percent level
** = both α_i and β_i significant

Although the data set (σ) contains somewhat more heterogeneous sales products of the industrial sectors than data set (π), the model (3.4) manages to capture a good deal of the variance in the material. R^2-values range from 0.42 to 0.97, although for the sector yielding the maximum degree of fit, the sample size is rather too small to warrant a closer inspection of that regression. Also in this case, model (3.4) is considerably superior to (3.1), which yields R^2-values in the range 0.08 – 0.62.

If we compare the distance 'elasticities' of Table 2 with those shown in Table 1, a somewhat scattered pattern may be noticed. Some industrial sectors reveal a larger distance attenuation of their total output sales than for their output of products intended for intermediary use, while for others the reverse holds. For more than half of the sectors, however, the distance elasticities are higher for sales of intermediaries than for total sales. This would indicate that for the majority of sectors, sales of investment and consumption goods (and services) are more regionally dispersed than those of intermediary products.

An overall observation concerning the production scale elasticities is that the $(\alpha_i + \beta_i)$–values are generally larger for the data set of total sales, indicating a greater need for intermediary shipments of investment (and consumption) goods on the margin. The sectors that exhibit the lowest elasticities remain the same when data set (σ) is used as for the data set (π).

The estimations presented in this section could serve as a basis for the construction of the a priori estimate of the interregional input-output table. Problems would then occur for some sectors for which the results were rather poor. On the other hand, the estimations above have an independent interest as tests of a common explanatory model type for shipment patterns. There are three main conclusions to be drawn from this analysis:

1. There exist significant differences in the distance attenuation of shipments of intermediaries between sectors which cannot be disregarded in a realistic attempt to construct an interregional input-output table.
2. There exist marked differences between the interregional delivery patterns of final demand products between sectors and also between final demand and intermediary products for a given sector.
3. There exist considerable differences between sectors in the relation between intermediary shipments and the production scales in the sending and receiving sector. For most sectors shipments depend almost linearly on the production scale, but exceptions exist.

Taken together, these conclusions motivate a continued data analysis, aimed at capturing some more of the individual traits of the shipment patterns to further ameliorate the a priori estimate of the interregional shipment table. This will be done below.

4.4.2 *Results of the contingency-table analysis*

In this presentation we will concentrate almost exclusively on the analysis of data set (π), i.e. the one containing information on intermediary shipments only. The main reason for this is that it is somewhat difficult to find a suitable form to represent the results in a compact way, making it necessary to select a few examples covering the main traits of the over-all results.

As noted above, see section 4.3.3, the estimations of the parameters of the Chenery-Moses model (3.5) have been performed by maximum-likelihood techniques. To solve the likelihood equations, a standard programme package, specially designed for estimation of log-linear models by proportionate fitting has been used.[4] As a measure of the goodness-of-fit between the model (3.5) and the data, a χ^2-measure has been used. The results of the estimations using the data set concerning intermediary shipments are presented in summary form in Table 3.

Table 3. Results of the estimations of model (3.5) using data set (π).

Sector	Degrees of freedom (df)	χ^2/df
(1) Agriculture and forestry	280	1593
(2) Mining and quarrying	392	176
(3) Food manufacturing	224	486
(4) Textile and clothing	560	282
(5) Wood and paper products	616	1657
(6) Printing and publishing	616	65
(7) Rubber products	392	148
(8) Chemical industry	672	877
(9) Petroleum and coal products	672	188
(10) Non-metallic mineral products	392	401
(11) Basic metal industries	504	1917
(12) Machinery and equipment	672	1190
(13) Shipbuilding	112	75
(14) Other manufacturing	224	183
(15) Electricity, gas, water, etc	672	68
(16) Construction	616	138
(17) Wholesale and retail trade	672	1982
(18) Transport and storage	672	90
(19) Other private services	672	282

[4] Ulf Nybert, Planning Office, Municipality of Stockholm, is thanked for computational assistance.

In judging the results of Table 3, it should be kept in mind that the data set consists only of rather large shipments so that in those positions of $x_{ij}^{kl}(\pi), i = 1, \ldots, m$, which contain non-zero elements, the registered transaction sums are often quite large. On the other hand many relations contain zero as the registered flow. To introduce a very simplified way the large number of small actual transactions which have not been registered in the data (in section 4.3.1 it was noted that the data covered some 2/3 of the intermediary deliveries) the following procedure has been adopted. For each sending sector a comparison with official statistics has revealed the coverage of the data of all transactions. The sum has been evenly distributed over all cells in the data set (without such a correction estimation problems would have occured using the current estimation technique).

In no case is the fit good enough to yield acceptance of the hypothesis of identical distributions using the χ^2-test. This is a typical result in cases where the sample size is very large. On the other hand, comparing the values of χ^2 divided by the number of degrees of freedom by sector with the regression results of Table 1, we see that the same over-all pattern appears. Those sectors which yield high values of R^2 for model (3.4) also tend to produce good fits in terms of χ^2 for model (3.5). Since the two models are quite similar in structure, this result is not surprising. It indicates, in fact, that the contingency-table approach in general leads to basically the same estimation results as the regression approach. The main difference is that the former captures somewhat more of the special traits of the data, being a markedly descriptive rather than explanatory approach, appropriate to generate 'smoothed' delivery data.

Although it is not the intention here to scrutinize the results of the estimations from a regional economic point of view, we will give a few glimpses of them in the form of a set of figures. In these figures the region effect, i.e. the factor B_i^{kl}, of equation (3.5) is depicted as a function of distance. The values are all normalized in the same way, i.e. so that the summed logarithmic effects are identical over both sending and receiving regions.

The idea behind the figures is that they should indicate the extent to which distance actually enters as an explanatory factor behind the region effect for different sectors, regions and product flows. The circled numbers in the figures correspond to the sending regions in terms of the notation introduced in Figure 2 (sec. 4.4).

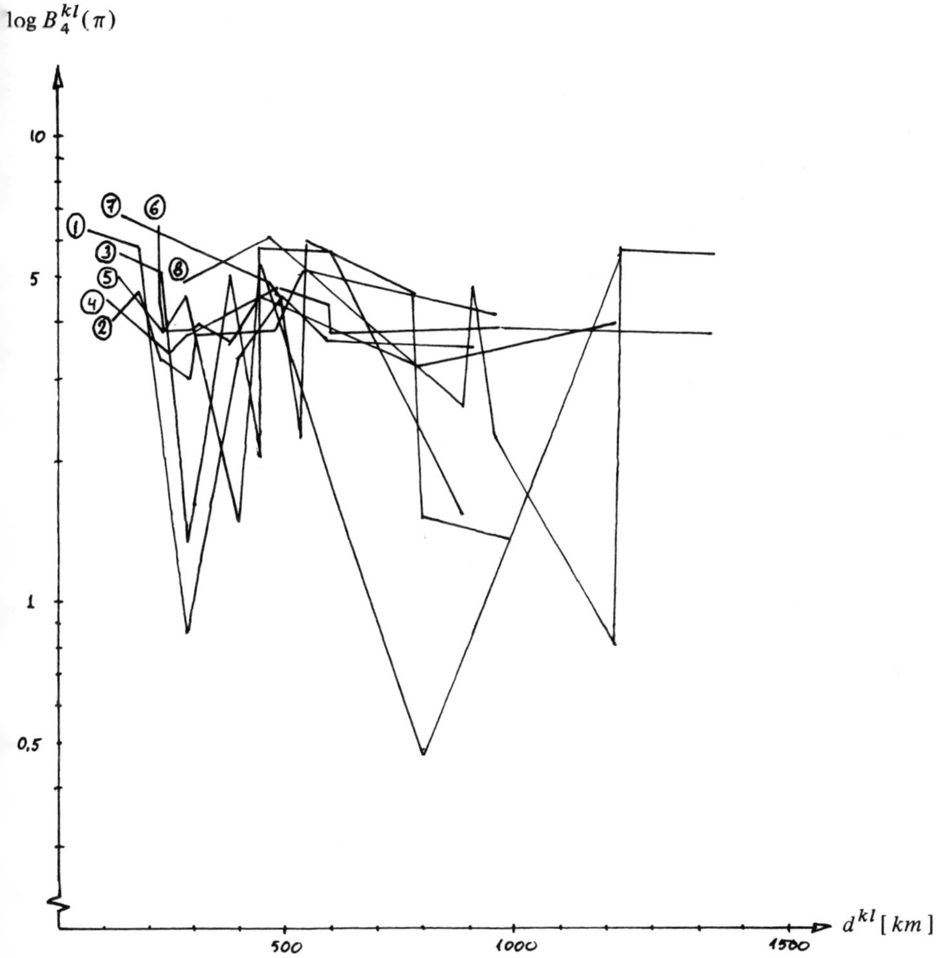

Fig. 3. Deliveries of intermediary products from sector 4 (textile and clothing). Region effect in logarithmic scale as a function of distance.

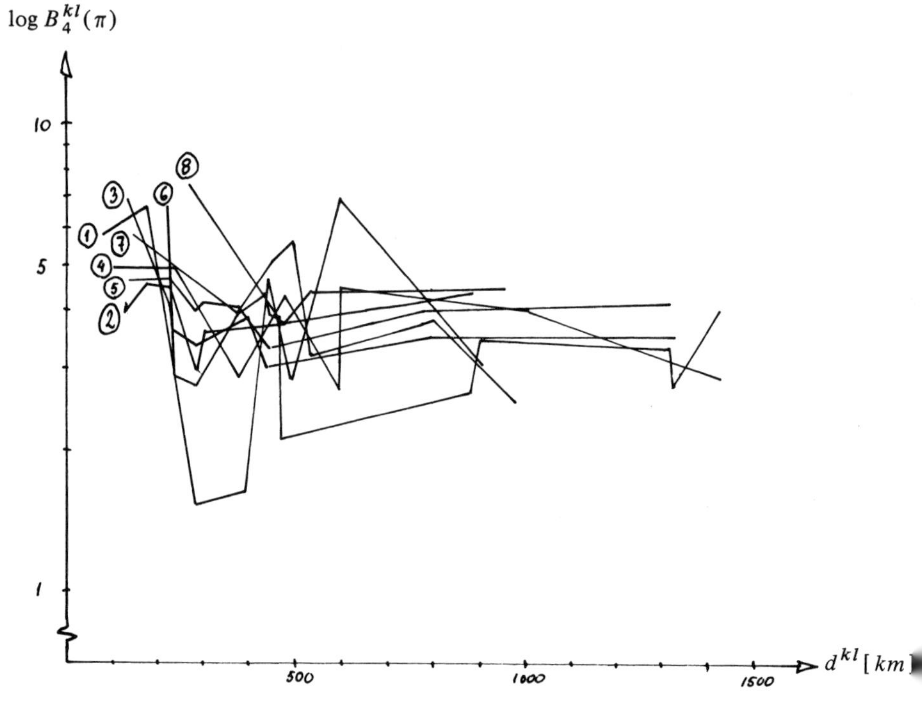

Fig. 4. Total sales of products from sector 4 (textile and clothing). Region effect in logarithmic scale as a function of distance.

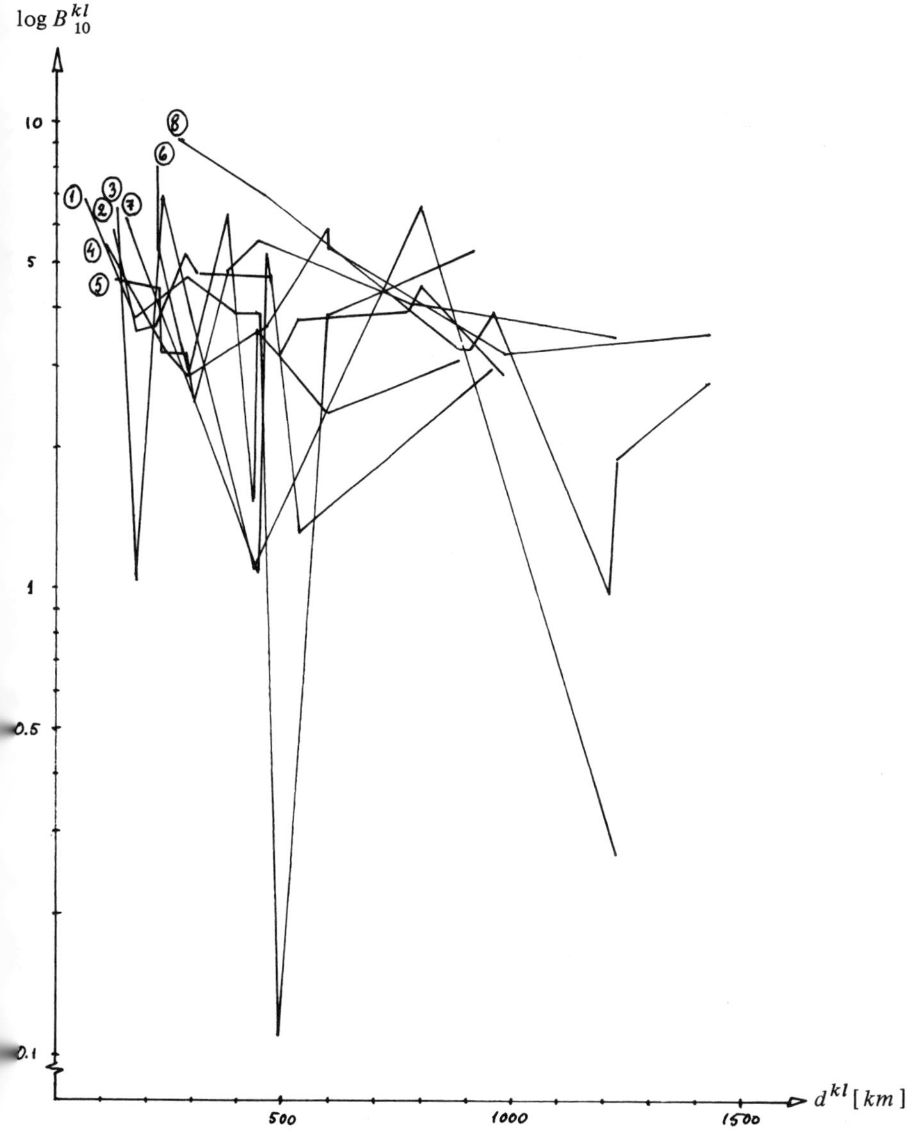

Fig. 5. Deliveries of intermediary products from sector 10 (non-metallic mineral products). Region effect in logarithmic scale as a function of distance.

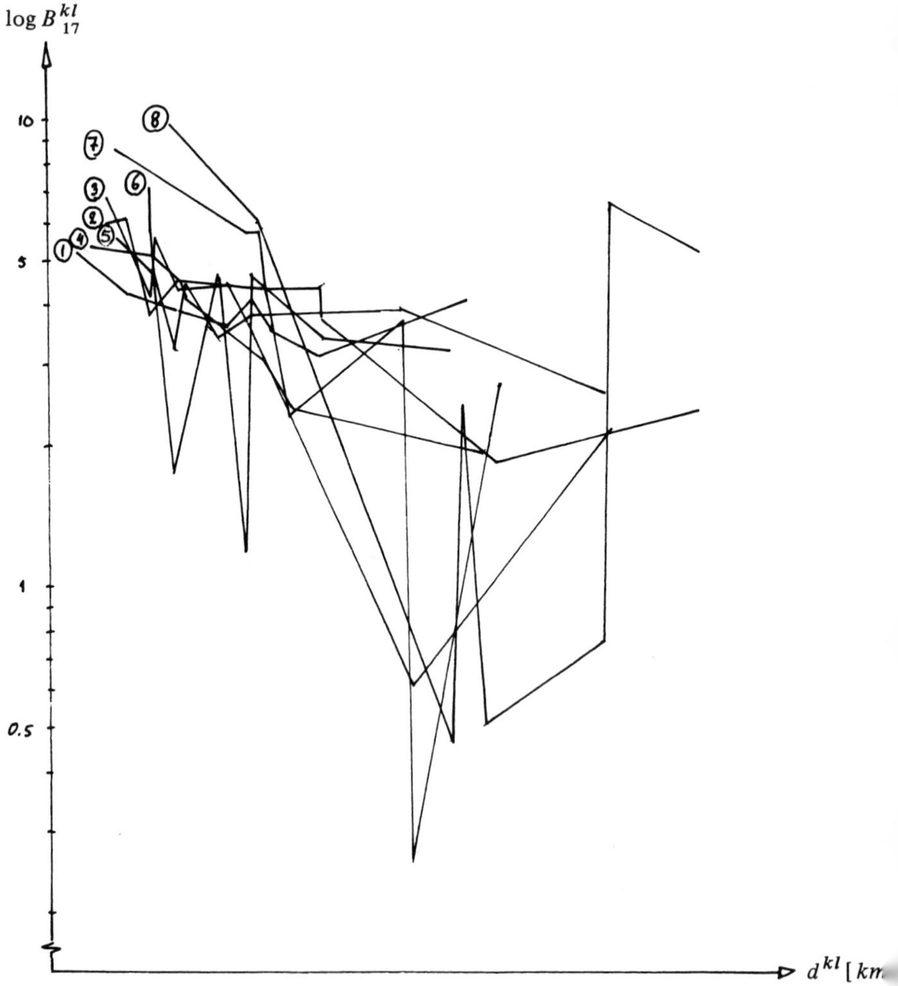

Fig. 6. Deliveries of intermediary products from sector 17 (wholesale and retail trade). Region effect in logarithmic scale as a function of distance.

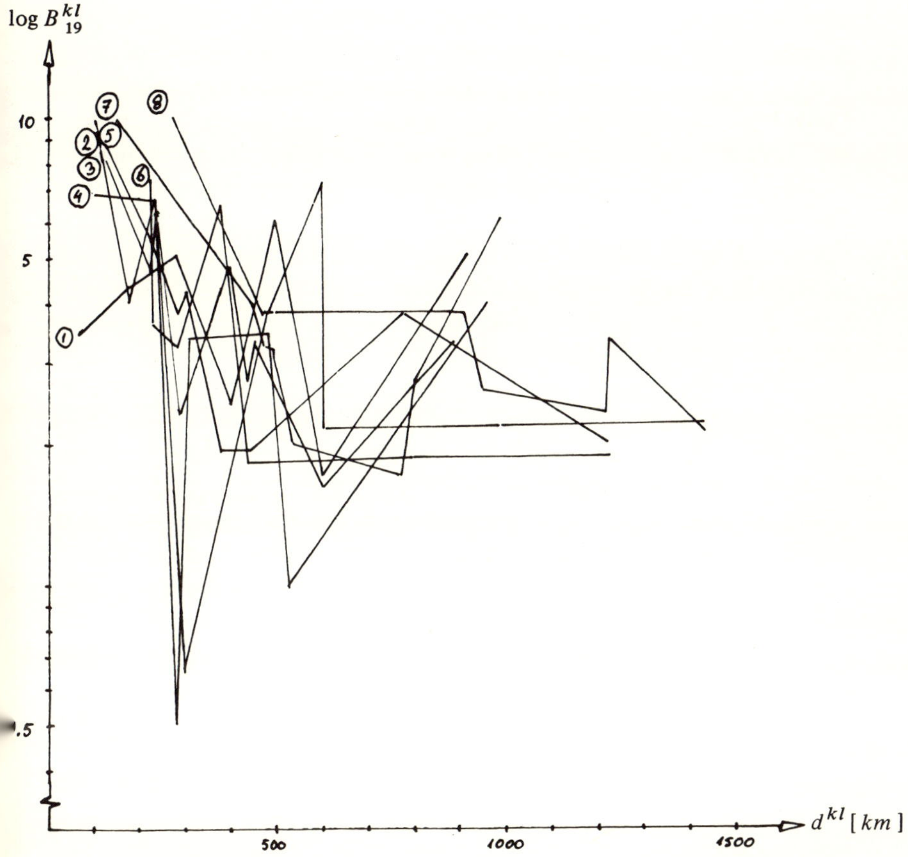

Fig. 7. Deliveries of intermediary products from sector 19 (other private services). Region effect in logarithmic scale as a function of distance.

The region effect values pertaining to each sending sector and region have been connected by straight lines. The more fluctuating these straight lines are, the less dominant is distance as an explanatory variable over others, such as firm structure and historical linkages.

The region effect for a common sector, viz. textile and clothing, but for varying data sets is depicted in Figures 3 and 4. A first observation is that distance seems to explain more of the pattern of total sales than of the pattern of intermediary deliveries. Thus final demand deliveries from this sector clearly are not traded in the same regional pattern as the intermediary ones. A further observation is that distance in effect seems to be a rather poor explanatory factor. There is a distance dependency in the patterns of both Figure 3 and 4 which actually very much resembles the one detected in the regressions above but the spread around this distance deterrence of shipment levels with distance is considerable. By transforming the figures into maps of Sweden a more explicitly geographical picture of the results may be given. We will not pursue this theme here, however.

From Figure 5, which gives the region effect for intermediary deliveries made by the sector non-metallic mineral products, we see that there may be large discrepancies between the distance dependency of shipments from different regions. For shipments originating in region 1 (Stockholm) no clear distance dependency can be discerned while shipments from region 8 (northernmost Sweden) exhibit a quite marked distance deterrence.

One unexpected result of the regression model estimations performed above was that the distance dependency of shipments made from the service sectors tended to be of the same magnitude as for some industrial sectors. Figures 6 and 7 shed some light on this question as regards two of the service sectors, viz. wholesale and retail trade and other private services (which contains, e.g. insurance, banking and consulting services). From Figure 6 we observe that the distance dependency varies systematrically between regions, as noted above. It is very weak for Stockholm but strong for northern parts of Sweden. This may be explained in terms of indivisibilities in the wholesale and retail trade, which gives the Stockholm region a very central position, receiving and delivering shipments irrespective of distance. In fact, the central position of Stockholm may also be explained by the fact that monetary shipments rather than physical shipment values are represented in the data.

Another interesting observation can be made in Figure 7 concerning the distance dependency of service inputs. It is evident for all regions but region 1 (Stockholm) that the distance deterrence of the deliveries originating from this sector does not fit an exponential model very well. The intraregional, short-distance shipments are quite large, whereas the interregional ones decline very slowly with distance. This indicates a heterogeneity in the product composition and conflicts with the results obtained in the regressions to a certain extent.

Although similar figures may be drawn for the technology effect, A^l_{ij}, we will not present any such results here. The technology effect will be adjusted by the exogenous data in the final estimations, but this will only to a very limited degree be the case with the distance effect. It is therefore very important for the reliability of the final result that the above 'smoothed' data describe *regional* product flows well. Due to the rather limited size of the current data set this may not always be the case here, however.

4.5 RESULTS OF THE MINIMUM INFORMATION ESTIMATIONS

In this section we will describe the results of the derivation of the complete interregional input-output table, using the mathematical programme (3.12) – (3.17). The aim of this presentation is not to give a detailed account of the results, their strengths and weaknesses, but to give an overview of the main traits of them, focussing once more on the spatial dimension.

In the estimations the a priori shipment matrix has been derived by the methodology described in section 4.3.3, using the estimates presented briefly in section 4.4.2. The focus in these estimations has been on the intermediary shipments, and the procedure described in section 4.3.4 for generating an a priori estimate of the patterns of shipment of finally demanded products has only been partly followed in this test of the methodology.

The solution of the mathematical programme (3.12) – (3.17) has been derived by application of the technique of interative proportionate fitting described earlier in section 4.4.2. Below, we give a brief summary of the final product of the estimations and present a simple application of the results using an aggregate version of the interregional table in order to demonstrate its applicability in dealing with interregional dependencies.

4.5.1 *Interregional linkages according to the results*

The current estimations make use of different sources of Swedish production data for the year 1975, whereas the delivery survey contains 1972 data. This means that there is a forecasting element contained in the analysis on top of the estimation one. The current study is intended to be a test of a methodological scheme which may be followed to get a (crude) interregional input-output table. Since the basic idea of this methodology is to make use of most of the various pieces of existing information concerning the interregional shipment pattern, there exists a real problem in finding data with which to compare the estimation results.

The current estimation framework has been set up to match the one used in the Swedish medium term economic forecasts, made by the Ministry of Economy (see Åberg, 1971, and official documents SOU 1976:42). There the transaction tables are measured in *purchaser's* prices. This means that transport and commercial cost margins have been included in the production values (of receiving firms). The main reason for this is that the consumption levels in the economic system are also measured in that way. This also means that total production levels are measured in terms of total input (see equations 2.1 – 2.12) where the various relations are summarized. However, these measurement problems are not central in the current analysis which may be applied also for an alternative treatment of transport and commercial margins, viz. by measuring production in terms of *producer's* prices. Since the only somewhat comprehensive Swedish data on intraregional input-output linkages that exists, viz. the ones collected by Roström and Sundberg (1978) (based on detailed survey information) are measured in producer's prices, there are only very limited possibilities of making comparisons.

Production levels and value added by region are supposed to be known in the current estimations as well as the intersectoral linkages and final demand levels at the national level. Final demand deliveries by sector *and* region as well as local area technical structures and interregional exports and imports are endogenous to the current estimation system. Since the focus in this presentation is on the spatial dimension, we will restrict the attention to a sectorally completely aggregated but regionally disaggregated system here. Such an interregional input-output table for Sweden 1975 according to the estimations is presented in Table 4 below. The regions are identified in Figure 2.

Table 4. Estimated interregional input-output table in purchaser's prices (million Skr in 1968 price level).

To / From	Intermediary inputs to regions 1	2	3	4	5	6	7	8	Sum of inter-mediary inputs	Domestic final demand	Export	Sum of final demand	Total input
1	10947	4556	1354	1795	2880	3047	2026	1628	28233	30856	7083	37939	66172
2	4812	11984	1526	2287	2563	1972	583	224	29951	21327	10010	31337	57288
3	1136	1382	6573	2033	1680	542	196	88	13630	13557	3794	17351	30981
4	4540	1492	2336	11737	2461	1013	142	250	23971	22961	6608	29569	53540
5	3371	2455	2197	2769	18559	2094	517	294	32256	31317	15441	46758	79014
6	953	1969	863	629	1889	7874	455	341	14973	12865	7736	20601	35574
7	156	1101	113	98	104	286	3396	399	5653	7023	3058	10081	15734
8	930	389	184	753	445	247	358	4171	7477	7761	3679	11440	18917
Sum of inter-mediary inputs	26845	25328	15146	22101	30581	17074	7673	7395	152144	94739	57409	205072	357220
Value added	29286	22500	12370	21424	27762	13589	6537	7703	134429				
Production (delivered prices)	56131	47828	27516	43525	58343	30663	14210	15098	293314				
Imports	10041	9460	3465	10015	20671	4911	1524	3819	63906				
Total Input	66172	57288	30981	53540	79014	35574	15734	18917	357220				

In Table 4 the interregional shipment matrix is determined in the estimations together with the column containing domestic final demand. The national totals are identical to those contained in the official Swedish medium term economic forecast made in 1975. The current matrix may be seen as an interregional transposition of the input-output table contained in that publication. It may be noted that the final demand deliveries are largest for region 1 (Stockholm) and for region 5 (western Sweden). The latter region has this property mainly because of its large export amounts. Another general observation is that there are remarkably strong interregional linkages in the economy, with region 1 (Stockholm) at the core of the national economy, exhibiting especially strong interregional ties.

4.5.2 *A simple interregional input-output model*

To illustrate further the interdependencies between regions in Sweden according to the estimations, the input-output table of Table 4 has been used to derive an interregional input-output model (see also Andersson, 1975). By invoking the linearity assumption of input patterns, we may derive input-output coefficients, a^{kl}, fulfilling the relation $a^{kl} = x^{kl}/x^l$. According to the results of section 4.4.2 such a linearity assumption may not be applied offhand to all sectors but here we use only the less restrictive version of it that contains total regional production levels, x^l, and aggregate interregional shipments of intermediaries, x^{kl}. In vector form we may rewrite the aggregate correspondence to (2.1) as follows, using the input-output model linearity assumption:

$$\boldsymbol{x} = \boldsymbol{A}\boldsymbol{x} + \boldsymbol{f} + \boldsymbol{e} - \boldsymbol{r} \tag{5.1}$$

If $(\boldsymbol{I}-\boldsymbol{A})$ is a non-negatively invertible matrix, which is usually the case (see, for example, Nikaido, 1968), (5.1) may of course be reexpressed to explain the production levels as functions of the various final demand components

$$\boldsymbol{x} = (\boldsymbol{I}-\boldsymbol{A})^{-1}\,(\boldsymbol{f} + \boldsymbol{e} - \boldsymbol{r}) \tag{5.2}$$

Expression (5.2) may be used to evaluate the total direct and indirect effects in the different regional economies, in reaction to changes in (one or several of the components of) final demand. The matrices $\boldsymbol{A}$ and $(\boldsymbol{I}-\boldsymbol{A})^{-1}$ are presented in Tables 5 and 6 for the current estimations.

Table 5. Interregional input-output matrix estimated by Swedish data 1975

From region \ To region	1	2	3	4	5	6	7	8
1	0.223	0.121	0.057	0.049	0.067	0.133	0.182	0.143
2	0.098	0.317	0.064	0.062	0.060	0.086	0.052	0.020
3	0.023	0.037	0.277	0.055	0.039	0.024	0.018	0.008
4	0.093	0.039	0.099	0.318	0.057	0.044	0.013	0.022
5	0.069	0.065	0.093	0.075	0.433	0.091	0.046	0.026
6	0.019	0.052	0.036	0.017	0.044	0.343	0.041	0.030
7	0.003	0.029	0.005	0.003	0.002	0.013	0.304	0.035
8	0.019	0.010	0.008	0.020	0.010	0.011	0.032	0.365
Sum	0.547	0.670	0.639	0.599	0.712	0.745	0.688	0.649

Table 6. Matrix of total direct and indirect effects on regional production levels of changes in final demand patterns.

From region \ To region	1	2	3	4	5	6	7	8	Sum
1	1.411	0.357	0.235	0.207	0.283	0.409	0.406	0.394	3.762
2	0.275	1.603	0.251	0.229	0.272	0.335	0.248	0.163	3.376
3	0.098	0.133	1.451	0.158	0.153	0.124	0.096	0.066	2.279
4	0.253	0.201	0.298	1.565	0.251	0.234	0.155	0.150	3.107
5	0.272	0.313	0.361	0.305	1.919	0.405	0.270	1.197	4.042
6	0.098	0.179	0.143	0.099	0.178	1.607	0.156	0.124	2.584
7	0.025	0.078	0.029	0.024	0.027	0.051	1.460	0.093	1.787
8	0.063	0.057	0.048	0.070	0.059	0.063	0.105	1.605	2.070
Sum	2.495	2.921	2.816	2.657	3.142	3.228	2.956	2.792	23.007

Table 5 reveals, among other things, that region 5 (western Sweden) is the most self-supplied one when it comes to intermediary deliveries, whereas region 1 (Stockholm) is the least self-supplied. The northern regions of Sweden (regions 7 and 8) exhibit relatively strong ties with the Stockholm area. From the column sums, which of course denote one minus the value added, we sec that the highest value added is to be found in Stockholm and the lowest in western Sweden with its marked international dependence.

The standard interpretation of columns in Table 6 is that they reveal the total production increase in all regions in reaction to an increase in the final demand levels in the various regions by one unit. An increase of final demand in region 6 (west central Sweden) is seen to have the largest *total* effect on the production levels, having a multiplier value well above 3.

The classical interpretation of rows in Table 6 is that their elements contain the total direct and indirect production effects for a certain region when the final demand in some region is raised marginally. If final demand is raised marginally throughout the system, the largest effect will be revealed in region 5 (western Sweden). The production multiplier is somewhat larger than 4. Here we will not transform these figures to employment multipliers since the presentation here is aimed at revealing the overall properties of the estimated table rather than answering specific planning and forecasting problems.

4.6 SOME CONCLUSIONS

A general methodology for estimating intersectoral and interregional input-output tables by efficient information-adding has been presented. The method has been thoroughly illustrated by performing a case-study, using Swedish survey and non-survey data. In the paper we have shown that it is actually possible to arrive at a consistent interregional table by making proper use of various pieces of information. However, it has not been a major goal to validate the method by making comparisons with other efforts in the same direction. Although there are attempts in Sweden to derive intraregional tables (see section 4.4.1), it has not been possible in this context to match definitions so as to obtain comparability. It is also worth stressing that the methodology is such that it is possible to exhaust practically all available pieces of information in the estimation procedure itself. This does not facilitate comparisons.

The data sets used in the paper calculate a priori estimates of delivery patterns are somewhat too unreliable to produce realistic estimates in all details. They contain only the larger shipments and these are measured in terms of money flows rather than in terms of values of physical flows. However, the basic idea is that it may well be realistic enough to use some kinds of survey data (samples of firms) to produce estimates of especially interregional linkages. In many applications the results are not taken ad notam but used as basically qualitative indicators of directions of change. Especially in a forecasting perspective it may not be unwise to use partial data in the way elaborated in the paper to produce somewhat crude, but still valuable forecasts of, e.g. labour market conditions in a regional system.

REFERENCES

Åberg, C. J. (1971), 'Plan and prognosis. A study of the methodology of the Swedish long term economic forecasts', Allmänna Förlaget, Stockholm (in Swedish).

Andersson, Å. E. (1975), 'A closed non-linear growth model for international and interregional trade and location', *Regional Science and Urban Economics*, 5, 427–444.

Bacharach, M. (1970), *Biproportionate matrices and input-output change*, Cambridge University Press, Cambridge.

Black, W. R. (1972), 'Interregional commodity flows: some experiments with the gravity model', *Journal of Regional Science*, 12, 107–118.

Chenerey, H. B. (1956), 'Interregional and international input-output analysis', In T. Barna (ed.), *The structural interdependence of the economy*, John Wiley and Sons, New York.

Gerking, S. D. (1976), *Estimation of stochastic input-output models*, Nijhoff, Leiden.

Goodman, L. (1963), 'Statistical methods for the preliminary analysis of transaction flows', *Econometrica*, 31, 197–208.

Gordon, I. R. (1976), 'Gravity demand functions, accessibility and regional trade', *Regional Studies*, 24–37.

Haberman, S. J. (1976), 'Recent development in the analysis of qualitative data', Technical Report 25, Department of Statistics, University of Chicago, Chicago.

Hårsman, B. and Marksjö, B. (1977), 'Modelling household changes by efficient information-adding', In Å. E. Andersson and I. Holmberg (eds.), *Demographic, economic and social interaction*, Ballinger, Boston.

Hartwick, J. M. (1971), 'Notes on the Isard and Chenery-Moses interregional input-output models', *Journal of Regional Science*, 11, 73–86.

Hobson, A. (1971), *Concepts in statistical mechanics*, Gordon and Breach, New York.

Ireland, C. T. and Scheuren, F. (1974), 'The RAKE's progress', Paper presented at the American Statistical Association Meeting, mimeo.

Jaynes, E. (1957), 'Information theory and statistical mechanics I', *Physical Review*, 106, 171–190.

de Kanter, J. and Morrison, W. I. (1976), 'The Merseyside input-output study and its application in structure planning', Paper presented at the Regional Science Association meeting, London.

Kullback, S. (1959), *Information theory and statistics*, John Wiley and Sons, New York.

Leontief, W. and Strout, A. (1966), 'Multiregional input-output analysis', In W. Leontief (ed.), *Input-output economics*, Oxford University Press, New York, 223-257.

Morrison, W. I. (1973), 'The development of an urban interindustry model 3: Input-output multipliers for Peterborough', *Environment and Planning*, 5, 545–554.

Moses, L. N. (1955), 'The stability of interregional trading patterns and input-output analysis', *American Economic Review*, 45, 803–832.

Nikaido, H. (1968), *Convex structures and economic theory*, Academic Press, New York.

Ohlsson, L. and Stighäll, C. (1976), 'The 1972 interregional delivery study. An assessment of field work, data quality and results', Departmentens Offsetcentral, Stockholm, (in Swedish).

O'Sullivan, P. and Ralston, B. (1974), 'Forecasting intercity commodity transport in the USA', *Regional Studies*, 8, 191–195.

Paelinck, J. and Nijkamp, P. (1976), *Operational theory and method in regional economics*, Saxon House Studies, London.

Richardson, H. W. (1972), *Input-output and regional economics*, Weidenfeld and Nicholson, London.

Riefler, R. (1973), 'Interregional input-output: a state of the arts survey', In Takayama, T. and Judge, G. G. (eds.), *Studies in economic planning over space and time*, North-Holland, Amsterdam.

Roström, B. and Sundberg, L. (1978), 'Documentation and technical description of input-output tables for Stockholm county', Technical report, Stockholm County Council Planning Office, Stockholm, (mimeo, in Swedish).

Savage, I. R. and Deutsch, K. W. (1960), 'A statistical model of the gross analysis of transaction flows', *Econometrica*, 18, 551–572.

Shannon, C. and Weaver, W. (1949), *'The mathematical theory of communication'*, University of Illinois Press, Urbana, Ill.

Snickars, F. and Weibull, J. (1977), 'A minimum information principle – theory and practice', *Regional Science and Urban Economics*, 7, 137–168.

SOU (1976), 'The model system of the medium and long term economic forecasting', Ministry of Economy, Stockholm, 42 (in Swedish).

Stilwell, F. J. B. and Boatright, B. D. (1971), 'A method of estimating interregional trade flows', *Regional and Urban Economics*, 1, 77–87.

5. Further evidence on alternative procedures for testing of spatial autocorrelation among regression disturbances

ANDRIES S. BRANDSMA and RONALD H. KETELLAPPER*

5.1 INTRODUCTION

It is well-known that the least-squares estimator of the vector of coefficients in a linear regression model is not Best-Linear Unbiased in the presence of autocorrelation. The problem was first recognized by statisticians dealing with temporal regression analysis. This led to the development of tests for temporal autocorrelation (see, for example, von Neumann, 1941; Durbin and Watson, 1950, 1951 and 1971). Because these tests require estimators for the unknown disturbances, attention was paid to the development of procedures to estimate the vector of disturbances. Apart from the obvious ordinary least squares (OLS) residuals we mention the best-linear unbiased scalar covariance (BLUS) estimator, which was proposed by Theil (1965) and Koerts (1967), and the best-linear unbiased fixed covariance (BLUF) residuals, which are described in Dubbelman, Abrahamse and Koerts (1972). Recently, Phillips and Harvey (1974) proposed a recursive procedure to obtain estimates of the errors. Because their estimator also has the LUS properties, we will refer to it as the RELUS-vector.

By absence of a natural ordering in space the concept of spatial autocorrelation is more general and complex than its analogue in time series analysis. This accounts for the separate treatment of the present topic, as for instance in Cliff and Ord (1972) and (1973) and Bartels and Hordijk (1977).

In this paper we aim at presenting some further results on the empirical power of alternative procedures for testing of spatial autocorrelation among disturbances in a cross-regional linear

---

*Econometric Institute, University of Groningen, P.O. Box 800, Groningen, the Netherlands. The authors are indebted to Cornelis P. A. Bartels for providing his experiences in the field and for suggesting improvements on the paper.

model. Recently, Bartels and Hordijk (1977) evaluated three test statistics. Using the coefficient proposed by Moran, the procedures differ in the estimator of the error vector being applied, viz. OLS, BLUS and RELUS.

The present paper adopts partly the same starting points; we consider the same classes of residuals, but we will give a more general and uniform representation of the first and second moment of the implied test statistics. Moreover, we study the likelihood ratio test as well as a test based on the maximum likelihood estimator for the coefficient of autocorrelation in the disturbances. Monte Carlo simulation experiments will be used to evaluate several characteristics of these test procedures, i.e. empirical power, robustness and appropriateness of distributional assumptions.

This paper is organized as follows. In section 5.2 the model and its assumptions are presented. Section 5.3 treats the first test statistic which we use, the Moran contiguity coefficient. In the same section we summarize the various residual estimation procedures which are employed. In section 5.4 the first, second and third moment of the Moran statistic are evaluated when the respective estimators for the disturbance vector are imputed. We generalize the formulae which have been developed for the OLS- and BLUS-cases to arbitrary linear unbiased estimators. As a result, more compact expressions are obtained. In section 5.5 the likelihood ratio test for the problem of spatial autocorrelation among regression disturbances is presented. The same section contains the definition of the so called naive test, a maximum likelihood based heuristic procedure. In section 5.6, the experimental design that we adopted is outlined, while section 5.7 presents the results. We conclude with some final suggestions.

5.2 FORMULATION OF THE STATISTICAL DECISION PROBLEM

The standard linear regression model in this study is

$$y = X\beta + \epsilon \tag{2.1}$$

where

y is the R-vector of observations on the regressand;
R is the number of spatial units that are considered;

X is the $R \times K$ matrix of observations on the K regressors;
$\boldsymbol{\beta}$ is the K-vector of regression coefficients and
$\boldsymbol{\epsilon}$ is the R-vector of unobservable disturbances.

We assume the regressors to be deterministic. Furthermore, we suppose that the errors are normally distributed and have zero expectations. The null hypothesis H_0 postulates independent homoscedastic errors

$$H_0 : \mathrm{E}(\boldsymbol{\epsilon}\, \boldsymbol{\epsilon}') = \sigma^2 \boldsymbol{I}_{(R)}, \tag{2.2}$$

where $\boldsymbol{I}_{(R)}$ is the $R \times R$ identity matrix. To formulate the alternative hypothesis H_1, we have to select a specific pattern of spatial dependencies, based on arguments from regional economics or geography. A complete analogy with the temporal case suggests a first-order spatial autocorrelation scheme with binary weights for spatial observations. However, within the present context one may prefer to allow for more differentiated weights, representing the supposed level of interaction between each pair of regions.

This leads to the formulation of the alternative hypothesis of first-order positive spatial autocorrelation

$$H_1 : \boldsymbol{\epsilon} = \rho \boldsymbol{W} \boldsymbol{\epsilon} + \boldsymbol{\zeta}, \quad \rho > 0 \tag{2.3}$$

where $\boldsymbol{\zeta} \sim N(\boldsymbol{0}, \sigma^2 \boldsymbol{I}_{(R)})$ and $\boldsymbol{W}$ is the matrix containing interactional weights, of which the diagonal elements are defined as zero. Following Cliff and Ord (1973, p. 137), we normalize the matrix by letting each row sum to unity. For reasons of simplicity we only consider the case of positive autocorrelation here. The procedures may easily be extended to other alternative hypotheses. This statistical decision problem may be written equivalently as

$$\begin{aligned}
&\boldsymbol{y} = X\boldsymbol{\beta} + \boldsymbol{\epsilon} \\
&\boldsymbol{\epsilon} \sim N(\boldsymbol{0}, \sigma^2 (\boldsymbol{I} - \rho \boldsymbol{W})^{-1} (\boldsymbol{I} - \rho \boldsymbol{W}')^{-1}) \\
&H_0 : \rho = 0 \\
&H_1 : \rho > 0
\end{aligned} \tag{2.4}$$

5.3 MORAN'S TEST STATISTIC

The first test statistic that we consider has been proposed originally as a measure of spatial autocorrelation of a general spatial series $z_1, \ldots, z_R$. Apart from a normalizing constant, Moran's statistic is defined as

$$MC = \frac{\sum_{i=1}^{R} \sum_{j=1}^{R} w_{ij} (z_i - \bar{z})(z_j - \bar{z})}{\sum_{j=1}^{R} (z_j - \bar{z})^2} \tag{3.1}$$

where $\bar{z}$ denotes the sample mean. As we use MC only as test-statistic, we deleted the constant. In order to use the statistic for testing of spatial autocorrelation among disturbances, one has to plug in an estimator for the error-vector.

The first disturbance estimater we consider is the OLS residual-vector. The OLS coefficient estimator being $\boldsymbol{b} = (\boldsymbol{X}'\boldsymbol{X})^{-1}\boldsymbol{X}'\boldsymbol{y}$, the residuals are defined by

$$°\boldsymbol{u} = \boldsymbol{y} - \boldsymbol{Xb} = (\boldsymbol{I} - \boldsymbol{X}(\boldsymbol{X}'\boldsymbol{X})^{-1}\boldsymbol{X}')\boldsymbol{y} = \boldsymbol{My} \tag{3.2}$$

Thus we conclude that the OLS residual is linear in the observations $\boldsymbol{y}$. Moreover, $\mathrm{E}(\boldsymbol{My}) = \mathrm{E}(\boldsymbol{MX\beta} + \boldsymbol{M\epsilon}) = \boldsymbol{0}$, as $\boldsymbol{MX} = \boldsymbol{0}$. This proves the residual's unbiasedness. In fact, under H_0 this estimator is best within the class of linear unbiased error-estimators (BLU) (see, for example, Theil, 1971). Contrary to the unobservable actual disturbances which are supposed to have a scalar covariance matrix, the covariance matrix of the OLS residual equals $\mathrm{E}(°\boldsymbol{u}°\boldsymbol{u}') = \sigma^2\boldsymbol{M}$. A drawback is the dependence of the moments of the resulting test-statistic on the observations on the regressors by means of the matrix $\boldsymbol{M}$.

The second class of disturbance estimators consists of vectors of the type $\boldsymbol{Ay}$, in which $\boldsymbol{A}$ is an $R \times R$ matrix, chosen in such a way that

1. the estimators are unbiased, i.e. $\mathrm{E}(\boldsymbol{Ay}) = \boldsymbol{0}$, equivalent with $\boldsymbol{AX} = \boldsymbol{0}$,
2. the covariance matrix is scalar, i.e. $\mathrm{E}(\boldsymbol{Ayy}'\boldsymbol{A}') = \sigma^2\boldsymbol{J}$, equivalent with $\boldsymbol{AA}' = \boldsymbol{J}$ under H_0, where

$$J = \begin{bmatrix} 0 & 0 \\ 0 & I_{(R-K)} \end{bmatrix}.$$

Condition (1) implies that A has a rank not greater than $(R-K)$, so at most $(R-K)$ residuals can be estimated independently. A specific element of this class of estimators is obtained by selecting a matrix A, which minimizes the expected sum of squared estimation errors $\mathrm{E}(\boldsymbol{Ay} - \boldsymbol{\epsilon})'(\boldsymbol{Ay} - \boldsymbol{\epsilon})$, under the restrictions (1) and (2). The solution of this problem yields the so-called best LUS (BLUS) estimators, cf. inter alia Theil (1971). A simple and more general treatment can be found in Dubbelman, Abrahamse and Koerts (1972) where the BLUS residuals are written as

$$^{B}\boldsymbol{u} = \boldsymbol{L}(\boldsymbol{L}'\boldsymbol{ML})^{-1/2}\boldsymbol{L}'\boldsymbol{My}, \text{ where } \boldsymbol{L} = \begin{bmatrix} 0 \\ I_{(R-K)} \end{bmatrix}. \tag{3.3}$$

The above solution is not unique, since there are $\binom{R}{K}$ possible ways in excluding K errors from estimating. When testing for autocorrelation in temporal models, it is intuitively clear that one should select successive observations to estimate and exclude observations at both ends of the time-series at hand. In that case there are at most $(K + 1)$ so-called bases, where a base consists of indices belonging to errors, excluded from estimation. In spatial analysis, however, it is more difficult to define 'boundary' regions. Dependent on the structure of the generalized contiguity matrix W one has to employ ad hoc techniques in choosing appropriate bases. In the sequel, we select the bases in such a way that the *number* of interactions that are excluded, is minimized. Theil (1971) recommends a procedure to select the base yielding the lowest expected sum of squares of estimation errors. As this criterion only reflects the performance of the estimator under H_0, the procedure does not necessarily select the base maximizing the power of the test against autocorrelation.

A second element of the LUS class of disturbance estimators which we will investigate here is the recursive LUS residual-vector (RELUS). Of course, it does not share the optimality conditions with OLS and BLUS estimators. Its appeal mainly consists of the relative ease of calculation and interpretation. According to Phillips and Harvey (1974), the RELUS vector may be obtained directly by

$$^{R}\boldsymbol{u} = C\boldsymbol{y} \tag{3.4}$$

where the i-th row of C is

$$\boldsymbol{c}_i' = (0, \ldots, 0) \quad , \qquad i = 1, \ldots, K$$

$$\boldsymbol{c}_i' = (-\boldsymbol{x}_i'(X_{i-1}'X_{i-1})^{-1}X_{i-1}/d_i, 1/d_i, 0, \ldots, 0), \quad i = K+1, \ldots, R,$$

in which $\boldsymbol{x}_i'$ is the i-th row of X, X_s and $\boldsymbol{y}_s$ are submatrices of X and $\boldsymbol{y}$ based on s observations and

$$d_i = (1 + \boldsymbol{x}_i'(X_{i-1}'X_{i-1})^{-1}\, \boldsymbol{x}_i)^{1/2} \quad , \quad i = K + 1, \ldots, R.$$

Note that in (3.3) and (3.4) the observations are rearranged such that the first K elements correspond to base-indices. Further on, the same convention is followed.

5.4 MOMENTS OF THE MORAN STATISTIC USING OLS AND LUS ESTIMATORS

In order to determine the critical regions of the various test statistics which are defined by substituting OLS, BLUS or RELUS residuals in Moran's coefficient, we have to know their distributions under H_0. Under normality of the true errors, there are procedures available to calculate the values of their distribution functions in arbitrary Points. For temporal models these are evaluated in Abrahamse and Koerts (1969) and L'Esperance and Taylor (1975). Dealing with spatial series, Bartels and Hordijk (1977) generalized the results. These exact densities could be used in order to find the significance points and to calculate the powers of different tests. However, this requires a great number of computations and rather lengthy computer programs. Alternatively, one may employ the normal distribution for testing with the Moran contiguity coefficient. In fact, this procedure has been followed in empirical applications up till now. This approximation is tolerable when the number of regions R is sufficiently large. For OLS residuals it has been proved that the asymptotic distribution of the Moran statistic is normal (cf. Cliff and Ord, 1973, p. 99). The same holds for the Moran statistic with BLUS or RELUS residuals, which can be proved as follows.

All three kinds of residuals we consider in this paper can be written as $\boldsymbol{u} = \boldsymbol{A}\,\boldsymbol{y}$. Moreover, all estimators are unbiased, so $\boldsymbol{AX} = \boldsymbol{0}$. This implies that the estimators are of the form

$$\boldsymbol{u} = \boldsymbol{A}(\boldsymbol{X\beta} + \boldsymbol{\epsilon}) = \boldsymbol{A}\,\boldsymbol{\epsilon} \tag{4.1}$$

The corresponding Moran test statistic can be written compactly as[1]

$$MC = \frac{\boldsymbol{u}'\boldsymbol{W}\boldsymbol{u}}{\boldsymbol{u}'\boldsymbol{u}} = \frac{\boldsymbol{\epsilon}'\boldsymbol{A}'\boldsymbol{W}\boldsymbol{A}\boldsymbol{\epsilon}}{\boldsymbol{\epsilon}'\boldsymbol{A}'\boldsymbol{A}\boldsymbol{\epsilon}} \tag{4.2}$$

Now the asymptotic normality depends on the eigenvalues of the matrix $\boldsymbol{A}'\boldsymbol{W}\boldsymbol{A}$, according to Cliff and Ord (1973, p. 99). However, the non-zero roots of $\boldsymbol{A}'\boldsymbol{W}\boldsymbol{A}$ are the same as the roots of the $(R-K) \times (R-K)$ submatrix of $\boldsymbol{W}$ in which columns and rows corresponding to base-elements are deleted. This proves the asymptotic normality of the Moran Coefficient for every LUS-vector. In order to evaluate the exact moments of MC, we need the following lemma, a modification of a theorem due to Pitman and Koopmans, which is formulated in Cliff and Ord (1973, p. 29):

- Let $z_1, \ldots, z_n$ be independently identically normally distributed variables: $\boldsymbol{z} = (z_1, \ldots, z_n)' \sim N(\boldsymbol{0}, \sigma^2 \boldsymbol{I})$.
- Every statistic $t(\boldsymbol{z})$ which is scale-free, that is $t(\lambda \boldsymbol{z}) = t(\boldsymbol{z})$, is independent of $s(\boldsymbol{z}) = \boldsymbol{z}_1' \boldsymbol{z}_1$, when $\boldsymbol{z}_1$ is any subvector of $\boldsymbol{z}$.

The proof is straightforward by calculating the joint characteristic function of t and s and using the theorem that this function equals the product of the separate characteristic functions if and only if t and s are independent (cf. Fisz, 1963, p. 123).

Before applying the lemma we note that for all three kinds of residuals holds that $\boldsymbol{A}'\boldsymbol{A} = \boldsymbol{M} = \boldsymbol{I} - \boldsymbol{X}(\boldsymbol{X}'\boldsymbol{X})^{-1}\boldsymbol{X}'$. For the OLS residual this is obvious as in the case $\boldsymbol{A} = \boldsymbol{M}$, while $\boldsymbol{M}$ is idempotent.

---

[1] Note that we do not subtract the sample mean of $\boldsymbol{u}$ as in (3.1). We think (4.2) is more appropriate since the population mean is known to be zero. At least, in the case of independent $N(0, \sigma^2)$ variables (as LUS residuals are), the estimator $R^{-1}\Sigma u_t^2$ has uniformly lower risk than $R^{-1}\Sigma(u_t - \bar{u})^2$. Moreover, in the case of OLS residuals, the sample mean is also zero in the presence of a constant term in $\boldsymbol{X}$, since $\boldsymbol{X}'\boldsymbol{M} = \boldsymbol{0}$.

The moments of MC, where the sample mean is subtracted, are derived in the Appendix.

Furthermore, every LUS estimator $\boldsymbol{C}y$ has the same length, i.e. $\boldsymbol{C}'\boldsymbol{C} = \boldsymbol{M}$, according to theorem 5.7 of Theil (1971).

Now we can rewrite (4.2) as

$$MC = \frac{\boldsymbol{\epsilon}'\boldsymbol{A}'\boldsymbol{W}\boldsymbol{A}\boldsymbol{\epsilon}}{\boldsymbol{\epsilon}'\boldsymbol{M}\boldsymbol{\epsilon}} \tag{4.3}$$

$\boldsymbol{M}$ is an idempotent matrix of rank $(R-K)$, which implies that it has $(R-K)$ roots equal to 1 and K roots equal to 0, so there exists an orthogonal matrix $\boldsymbol{P}$ such that

$$\boldsymbol{P}'\boldsymbol{M}\boldsymbol{P} = \begin{bmatrix} \boldsymbol{I}_{(R-K)} & \boldsymbol{0} \\ \boldsymbol{0} & \boldsymbol{0} \end{bmatrix}$$

$\boldsymbol{P}$ is the transpose of any orthogonal set of eigenvectors. Defining $\boldsymbol{\nu} = \boldsymbol{P}'\boldsymbol{\epsilon}$ gives $\boldsymbol{\nu} \sim N(\boldsymbol{0}, \sigma^2 \boldsymbol{I})$; now

$$\boldsymbol{\epsilon}'\boldsymbol{M}\boldsymbol{\epsilon} = \boldsymbol{\epsilon}'\boldsymbol{P}\begin{bmatrix} \boldsymbol{I} & \boldsymbol{0} \\ \boldsymbol{0} & \boldsymbol{0} \end{bmatrix}\boldsymbol{P}'\boldsymbol{\epsilon} = \boldsymbol{\nu}_1'\boldsymbol{\nu}_1,$$

where $\boldsymbol{\nu}_1$ is a $(R-K)$ subvector of $\boldsymbol{\nu}$.

Furthermore, $\boldsymbol{\epsilon}'\boldsymbol{A}'\boldsymbol{W}\boldsymbol{A}\boldsymbol{\epsilon}$ is also a quadratic form in $\boldsymbol{\nu}$:

$$\boldsymbol{\epsilon}'\boldsymbol{A}'\boldsymbol{W}\boldsymbol{A}\boldsymbol{\epsilon} = \boldsymbol{\nu}'\boldsymbol{P}'\boldsymbol{A}'\boldsymbol{W}\boldsymbol{A}\boldsymbol{P}\boldsymbol{\nu}$$

This enables application of the Pitman-Koopmans lemma, stating that MC and $\boldsymbol{\epsilon}'\boldsymbol{M}\boldsymbol{\epsilon}$ are independent; hence:

$$\begin{aligned} \mathrm{E}(\boldsymbol{\epsilon}'\boldsymbol{A}'\boldsymbol{W}\boldsymbol{A}\boldsymbol{\epsilon})^p &= \mathrm{E}((MC)^p \,.\, (\boldsymbol{\epsilon}'\boldsymbol{M}\boldsymbol{\epsilon})^p) \\ &= \mathrm{E}\,(MC)^p \,.\, \mathrm{E}(\boldsymbol{\epsilon}'\boldsymbol{M}\boldsymbol{\epsilon})^p, \end{aligned}$$

which yields

$$\mathrm{E}\,(MC)^p = \frac{\mathrm{E}(\boldsymbol{\epsilon}'\boldsymbol{A}'\boldsymbol{W}\boldsymbol{A}\boldsymbol{\epsilon})^p}{\mathrm{E}(\boldsymbol{\epsilon}'\boldsymbol{M}\boldsymbol{\epsilon})^p} \tag{4.4}$$

This result can be used to calculate the first two moments of MC. We have $\mathrm{E}(\boldsymbol{\epsilon}'\boldsymbol{A}'\boldsymbol{W}\boldsymbol{A}\boldsymbol{\epsilon}) = \mathrm{E}(tr(\boldsymbol{\epsilon}'\boldsymbol{A}'\boldsymbol{W}\boldsymbol{A}\boldsymbol{\epsilon})) = \mathrm{E}(tr(\boldsymbol{A}'\boldsymbol{W}\boldsymbol{A}\boldsymbol{\epsilon}\boldsymbol{\epsilon}')) = \sigma^2 tr(\boldsymbol{A}'\boldsymbol{W}\boldsymbol{A})$, where tr stands for the trace-operator. Also,

$E(\boldsymbol{\epsilon}'\boldsymbol{M}\boldsymbol{\epsilon}) = \sigma^2 tr(\boldsymbol{M}) = \sigma^2 (R-K)$ which gives the general first moment

$$E(MC) = \frac{tr(\boldsymbol{A}'\boldsymbol{W}\boldsymbol{A})}{R-K} \tag{4.5}$$

Furthermore, $E(\boldsymbol{\epsilon}'\boldsymbol{M}\boldsymbol{\epsilon})^2 = \sigma^4 (R-K)(R-K+2)$, because

$$\frac{\boldsymbol{\epsilon}'\boldsymbol{M}\boldsymbol{\epsilon}}{\sigma^2} \sim \chi^2_{R-K}$$

More general, we have

$$\begin{aligned} E(\boldsymbol{\epsilon}'\boldsymbol{B}\boldsymbol{\epsilon})^2 &= E(\sum_{i \neq j}\sum b_{ii} b_{jj} \epsilon_i^2 \epsilon_j^2 + \sum_{i \neq j}\sum b_{ij} b_{ji} \epsilon_i^2 \epsilon_j^2 \\ &+ \sum_{i \neq j}\sum b_{ij}^2 \epsilon_i^2 \epsilon_j^2 + \sum_i b_{ii}^2 \epsilon_i^4) \qquad (4.6) \\ &= \sigma^4 (tr(\boldsymbol{B}\boldsymbol{B}') + tr(\boldsymbol{B}^2) + (tr\boldsymbol{B})^2) \end{aligned}$$

Substituting $\boldsymbol{B} = \boldsymbol{A}'\boldsymbol{W}\boldsymbol{A}$ finally gives the general second moment of the numerator of the Moran statistic. Note that these formulae do not only hold for OLS, BLUS and RELUS estimators, but also for the complete class of LUS residuals. Below, the procedures are specialized to the OLS and LUS cases.

1. OLS-case:

The Moran coefficient reduces to

$$MC = \frac{\boldsymbol{\epsilon}'\boldsymbol{M}\boldsymbol{W}\boldsymbol{M}\boldsymbol{\epsilon}}{\boldsymbol{\epsilon}'\boldsymbol{M}\boldsymbol{\epsilon}}, \text{ so}$$

$$E(MC) = \frac{tr(\boldsymbol{M}\boldsymbol{W}\boldsymbol{M})}{R-K} = \frac{tr(\boldsymbol{M}\boldsymbol{W})}{R-K} \tag{4.7}$$

Furthermore,

$$E(MC^2) = \frac{tr(\boldsymbol{M}\boldsymbol{W}\boldsymbol{M}\boldsymbol{W}') + tr((\boldsymbol{M}\boldsymbol{W})^2) + (tr(\boldsymbol{M}\boldsymbol{W}))^2}{(R-K)(R-K+2)} \tag{4.8}$$

2. LUS-cases:

All LUS residuals are distributed $N(\mathbf{0}, \sigma^2 \mathbf{J})$, so corresponding moments of different residuals estimators are equal ($\mathbf{J}$ defined as before). The moments of MC may be rewritten as

$$\mathrm{E}(MC) = \frac{tr(\mathbf{A}'\mathbf{W}\mathbf{A})}{R-K} = \frac{tr(\mathbf{W}\mathbf{J})}{R-K} = 0 \tag{4.9}$$

Furthermore,

$$\mathrm{E}(MC^2) = \frac{tr(\mathbf{W}_0\mathbf{W}_0') + tr((\mathbf{W}_0)^2)}{(R-K)(R-K+2)}, \tag{4.10}$$

where $\mathbf{W}_0$ is the $(R-K) \times (R-K)$ submatrix of $\mathbf{W}$ in which rows and columns, corresponding to base elements are deleted.

NOTE: The third moment of MC can be obtained along similar lines. This will be used in Section 6 to calculate the skewness of the distribution of MC, given $\mathbf{W}$, $\mathbf{X}$ and a particular base in case of LUS residuals. The selected coefficient of skewness is defined by

$$S = \frac{\mathrm{E}(MC - \mathrm{E}(MC))^3}{(\mathrm{VAR}(MC))^{3/2}} \tag{4.11}$$

5.5 THE LIKELIHOOD RATIO TEST

In the preceding section we followed a two-step procedure in order to arrive at a test-statistic. The first step consists of estimating the unknown errors. Secondly, these estimates are substituted in the Moran coefficient. In this section, we consider a simultaneous approach in testing for autocorrelation among regression disturbances. As a by-product of this approach, the maximum likelihood estimator for ρ is obtained. Thus we derive two tests, namely the likelihood ratio test and a test based on the maximum likelihood estimator $\hat{\rho}$.

The joint density of $y_1, \ldots, y_R$ can be written as

$$f(y_1, y_2, \ldots, y_R; \boldsymbol{\beta}, \sigma^2, \rho) = (2\pi)^{-R/2} \sigma^{-R} \, |V|^{-1/2} \exp\{-\frac{1}{2\sigma^2} (y-X\boldsymbol{\beta})' V^{-1} (y-X\boldsymbol{\beta})\}, \tag{5.1}$$

where $V = (I-\rho W)^{-1} (I-\rho W')^{-1}$, the covariance matrix of the errors $\boldsymbol{\epsilon}$. The outcome of the LR test statistic for the problem of testing for autocorrelation (including negative autocorrelation for the time being) is defined by

$$l = \frac{\max\limits_{\boldsymbol{\beta}, \sigma^2} f(y_1, \ldots, y_R; \boldsymbol{\beta}, \sigma^2, 0)}{\max\limits_{\boldsymbol{\beta}, \sigma^2, \rho} f(y_1, \ldots, y_R; \boldsymbol{\beta}, \sigma^2, \rho)} \tag{5.2}$$

When the null-hypothesis is true, $-2 \log l$ is the outcome of an asymptotically χ_1^2 distributed variable under regularity conditions (cf. Kendall and Stuart 1973, section 24-7). We use this distribution to determine the critical region of the LR test (i.e. the interval 3.841, ∞) when the significance level is 5%).

Evaluation of the numerator of (5.2) is simple, since the maximum likelihood solution for $\boldsymbol{\beta}$ equals the last-squares estimate under H_0: $b = (X'X)^{-1}X'y$, while $\hat{\sigma}^2 = \frac{1}{R}(y-Xb)'(y-Xb)$. For the denumerator no analytically expression is available, however. The log-likelihood corresponding to the density (5.1) is

$$L = -\frac{R}{2}\log(2\pi) - \frac{R}{2}\log\sigma^2 - \frac{1}{2}\log|V| - \frac{1}{2\sigma^2}(y-X\boldsymbol{\beta})'V^{-1}(y-X\boldsymbol{\beta}) \tag{5.3}$$

Differentiating (5.3) with respect to the parameters and setting them equal to 0 yields (see also Hepple, 1976):

$$1.\quad \frac{\partial L}{\partial \sigma^2} = -\frac{R}{2\sigma^2} + \frac{1}{2\sigma^4}(y-X\boldsymbol{\beta})'V^{-1}(y-X\boldsymbol{\beta}) = 0$$

$$2.\quad \frac{\partial L}{\partial \boldsymbol{\beta}} = \frac{1}{\sigma^2} X'V^{-1}(y-X\boldsymbol{\beta}) = \mathbf{0} \tag{5.4}$$

$$3.\quad \frac{\partial L}{\partial \rho} = -tr(WP^{-1}) + \frac{1}{\sigma^2}(y-X\boldsymbol{\beta})'P'W(y-X\boldsymbol{\beta}) = 0\,,$$

where $P = (I - \rho W)$.

The necessary conditions imply that the ML estimate $\hat{\boldsymbol{\beta}}$ is the Aitken estimate after substitution of the ML estimate $\hat{\rho}$ in V:

$$\hat{\boldsymbol{\beta}} = (X'V^{-1}X)^{-1}X'V^{-1}y, \text{ while} \tag{5.5}$$

$$\hat{\sigma}^2 = \frac{1}{R}(y-X\hat{\boldsymbol{\beta}})'V^{-1}(y-X\hat{\boldsymbol{\beta}}) \tag{5.6}$$

Substitution of these formulae in (5.4), 3 gives a nonlinear equation in ρ, the roots of which produce the stationary points of the maximization problem.

One may also proceed by maximizing the likelihood by employing steepest-descent or gradient optimization algorithms. In this case it is computationally advantageous to substitute $\hat{\boldsymbol{\beta}}$ and $\hat{\sigma}^2$, so that the likelihood is only dependent on ρ.

To obtain a one-sided test, we modified the LR test in such a way that only outcomes of the LR test statistic, belonging to positive ML estimates for ρ, are able to fall in the critical region. The critical region becomes in this case 2.706, ∞ (significance level 5%).

Finally we study a LR similar test, to be called the naive test. In other applications, as for instance simultaneous linear econometric models, it is customary to consider an estimated parameter as significant, when its absolute value is greater than twice its estimated standard deviation. In fact, this two-sided test is based on the normal distribution, the 97.5-th percentile of which is 1.96. Analogously, we define the outcome of the NT statistic as:

$$NT = \hat{\rho}\,/(\hat{\text{var}}(\hat{\rho}))^{1/2}, \text{ where} \tag{5.7}$$

$\hat{\rho}$ is the ML estimate of ρ and $\hat{\text{var}}(\hat{\rho})$ is the estimate of the variance of $\hat{\rho}$, obtained from the Cramér-Rao bound in the point $\rho = \hat{\rho}$. Since we require a one-sided test, the critical region is 1.645, ∞.

5.6 SIMULATION STUDY OF THE MORAN AND LIKELIHOOD RATIO TESTS

The performance of the five tests, presented in sections 3 and 5, was studied by examining four characteristics: (1) the empirical distribution of the test-statistic under H_0, (2) the number of outcomes in the critical region under H_0, (3) the empirical power and (4) the sensitivity to the normality-assumption. The first two characteristics are considered in order to check whether the normal distribution in case of the Moran test and the naive test and the χ^2 distribution in case of the likelihood ratio test are appropriate. Characteristic (3) should select the best statistic, at least if the performance on (1) and (2) is satisfactory. The sensitivity analysis, finally, is supposed to reveal which tests are robust.

The simulation experiments were designed as follows. Together with two different spatial structures of the Netherlands, we had 4 regressor sets at our disposal, obtained from previous empirical studies. Given a certain X matrix, for each ρ – having values of resp. 0.0, 0.1, 0.5 and 0.9 – 100 regressand vectors were simulated by generating autocorrelated errors according to the scheme of Section 2 and by fixing a coefficient vector. Then the outcomes of the various test statistics are calculated (the Moran statistic was standardized with the corresponding exact moments) and compared with the critical values of the tests. The empirical power is now defined as the percentage of outcomes in the critical region. Formally the power is only defined for parameter values belonging to H_1. The row $\rho = 0$ in the tables also gives the % of the outcomes in the critical region and may be interpreted as the empirical significance level. Theoretically this value should be 5%.

In order to judge the performance with respect to characteristic (1), we plotted a frequency polygon of the standardized outcomes of each Moran statistic and of the NT–statistic. Concerning the LR test-statistic, we plotted a frequency polygon of sign $(\hat{\rho}).\sqrt{l}$, where $\hat{\rho}$ is the ML estimate of ρ and l is the outcome of the test (5.2). This plotted statistic should be $N(0, 1)$–distributed, since l is χ_1^2-distributed under H_0. The standard normal densities were plotted in the same graphs (dotted lines). The question whether the empirical density fits the theoretical density was tested with Pearson's chi-square test; compare, for example, Koerts and Abrahamse (1971). The successive boundary points of the intervals are: −1.5, −1, −0.5, 0, 0.5, 1, 1.5. The corresponding theoretical frequencies are 7, 9, 15, 19, 19, 15, 9 and 7. In this case Pearson's chi-square test has 7 degrees of freedom; its critical region is 14.1, ∞.

An indication about the robustness of the tests was obtained by generating errors such that the density of the ζ's is exponential rather than normal. In fact we used the density

$$f(x) = \tfrac{1}{2}\,\lambda e^{-\lambda |x|},$$

where λ is chosen as $\sqrt{2}/\sigma^2$ in order to obtain a density with variance equal to σ^2. We used this density because of its 'large tail', i.e. the kurtosis is 3. This density, together with the $N(0, \sigma^2)$ density is plotted in Fig. 1 ($\sigma = 1$).

5.7 RESULTS

In this section we present the results of applying the 5 test-procedures to 4 different linear models, differing on features which are expected to affect the results. The first 3 models correspond with examples 1, 2 and 3 in Bartels and Hordijk (1977). As the result of an error in their computer program and because of their alternative definition of *MC*, the values of the powers as calculated by them differ from the results presented here. The outcomes of Pearson's (χ^2)-test and the skewness of the distribution(s) (only under normality) are given in the last two rows of Table 1 until 8.

We only present the plots of the empirical distributions for the Moran-statistic with OLS-residuals and for the transformed LR-statistic. The cases of LUS-residuals were deleted because they behaved in a similar way as OLS, at least concerning this criterion. The *NT* behaves as equally bad as the LR-test. The correlation matrices of the explanatory variables of each model are given in Table 9.

MODEL 1. The Netherlands, 39 regions, 3 variables.
In this first model we use data from a study of differences in level

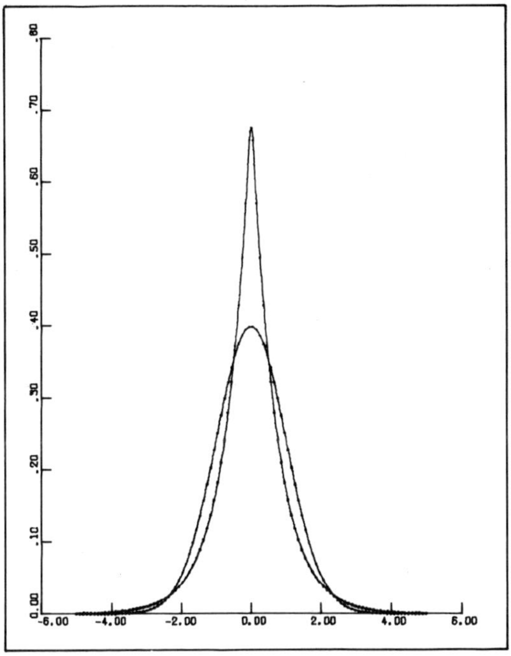

Fig. 1. Normal and exponential densities

Fig. 2. 40 Dutch COROP regions

and inequality of personal incomes by Bartels and Bertens (1976). They used a regionalization of the Netherlands into 40 so-called COROP regions (Fig. 2). The regressors consist of a constant term, the average number of years of schooling and the percentage of labour force with monopolistic power at the labour market. The coefficient vector, used as input in the generation of the regressand, is (0.1, 1.0, 0.5)′. Three bases are compared, namely those formed by the regions 2, 31 and 39; 24, 31 and 39; 27, 31 and 39. The results on the empirical powers are presented in Table 1 and 2, referring to the cases where normal respectively exponential errors are generated.

Table 1. Powers of 5 test procedures for Model 1, $R = 39, K = 3$ (normal error density)

ρ	OLS	BLUS 1	BLUS 2	BLUS 3	RELUS 1	RELUS 2	RELUS 3	LR	NT
0.0	7	5	7	7	6	6	4	3	3
0.1	9	6	5	6	7	6	4	1	6
0.5	79	65	63	60	58	61	50	67	73
0.9	100	97	99	96	94	98	97	100	100
χ^2	7.1	12.3	12.8	11.2	3.1	3.0	7.5	11.3	17.3
S	0.33	0.26	0.26	0.26	0.26	0.26	0.26	–	–

Table 2. Powers of 5 test procedures for Model 1, $R = 39, K = 3$ (exponential error density)

ρ	OLS	BLUS 1	BLUS 2	BLUS 3	RELUS 1	RELUS 2	RELUS 3	LR	NT
0.0	4	6	5	5	6	3	4	0	1
0.1	11	14	18	17	13	18	15	6	10
0.5	66	56	63	63	55	55	54	50	66
0.9	100	99	100	99	98	100	98	100	100
χ^2	2.8	8.1	4.1	5.1	7.2	1.0	9.8	41.5	55.9

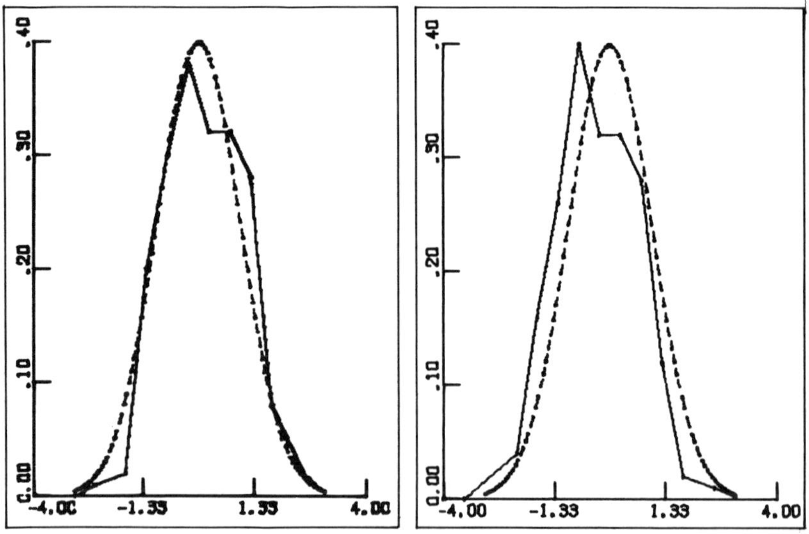

OLS ($\chi^2 = 7.1, S = 0.33$) Likelihood ratio ($\chi^2 = 11.3$)

Fig. 3. Empirical densities of 2 test-procedures (Model 1)

MODEL 2. The Netherlands, 39 regions, 6 variables.
The regressor set of model 1 is extended with the following three variables: percentage of labour force in manufacturing, participation rate and unemployment rate. The vector of coefficients we used now was (0.1, 1.0, 0.5, 0.2, 0.1, –0.2)′. The four distinct bases we considered are constituted by regions 2, 4, 24, 27, 31, 39; 2, 12, 24, 27, 31, 39; 2, 19, 24, 27, 31, 39 and 2, 24, 27, 31, 38, 39.

Table 3. Powers of 5 test procedures for Model 2, $R = 39, K = 6$ (normal error density)

ρ	OLS	BLUS 1	BLUS 2	BLUS 3	BLUS 4	RELUS 1	RELUS 2	RELUS 3	RELUS 4	LR	NT
0.0	5	6	5	4	7	7	10	7	8	5	6
0.1	10	5	6	5	4	5	4	7	3	1	10
0.5	63	41	43	44	42	43	39	38	40	50	63
0.9	98	84	83	86	89	86	86	86	86	99	100
χ^2	1.6	2.5	3.5	6.3	1.6	6.6	4.9	7.6	5.9	44.0	95.9
S	0.41	0.28	0.28	0.28	0.28	0.28	0.28	0.30	0.28	–	–

Table 4. Powers of 5 test procedures for Model 2, $R = 39, K = 6$ (exponential error density)

ρ	OLS	BLUS 1	BLUS 2	BLUS 3	BLUS 4	RELUS 1	RELUS 2	RELUS 3	RELUS 4	LR	NT
0.0	4	4	6	4	4	5	5	3	3	0	2
0.1	10	11	12	11	9	16	14	14	13	5	9
0.5	54	43	41	44	46	41	36	33	35	39	57
0.9	99	83	83	87	88	90	90	84	91	100	100
χ^2	6.6	8.5	7.1	6.5	8.9	4.0	5.2	7.6	8.7	67.6	113.6

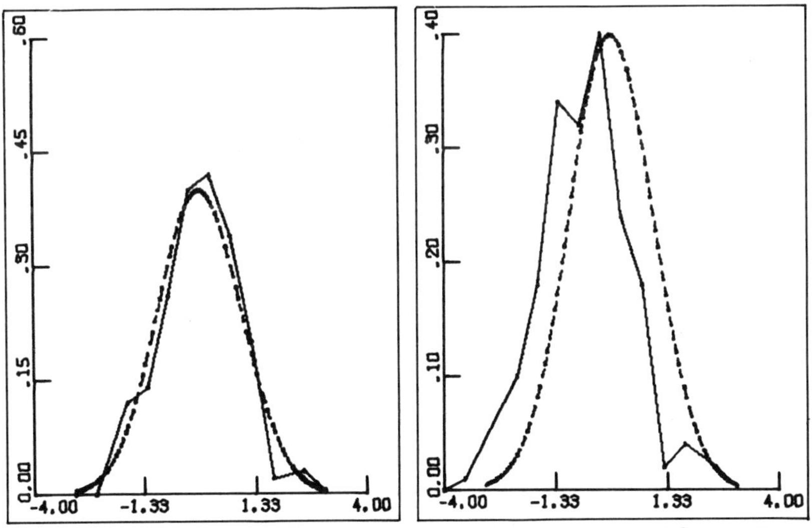

OLS (χ^2 = 1.6, S = 0.41) Likelihood ratio (χ^2 = 44.0)

Fig. 4. Empirical densities of 2 test-procedures (Model 2)

MODEL 3. The Netherlands, 39 regions, 3 variables.
Based on the same spatial structure as in the two previous models, we now obtained two regressors by generating spatial autocorrelated series. For the second regressor an autocorrelation coefficient of 0.8 was used, for the third 0.2. The drawings are standard normally distributed. The first regressor is the constant term.

Table 5. Powers of 5 test procedures for Model 3, R = 39, K = 3 (normal error density)

ρ	OLS	BLUS 1	BLUS 2	BLUS 3	RELUS 1	RELUS 2	RELUS 3	LR	NT
0.0	7	8	5	5	7	6	5	4	4
0.1	8	7	6	9	12	8	11	4	7
0.5	85	70	73	73	67	66	64	75	79
0.9	100	99	99	97	98	96	96	100	100
χ^2	17.4	3.7	2.5	6.0	7.6	9.9	7.0	10.3	17.1
S	0.31	0.26	0.26	0.26	0.26	0.26	0.26	–	–

Table 6. Powers of 5 test procedures for Model 3, R = 39, K = 3 (exponential error density)

ρ	OLS	BLUS 1	BLUS 2	BLUS 3	RELUS 1	RELUS 2	RELUS 3	LR	NT
0.0	6	5	3	4	5	4	4	0	2
0.1	5	7	7	9	7	7	7	4	7
0.5	67	58	59	57	54	52	53	61	69
0.9	99	98	98	98	98	95	98	100	100
χ^2	15.1	15.9	26.5	20.8	12.3	11.3	7.5	20.8	44.8

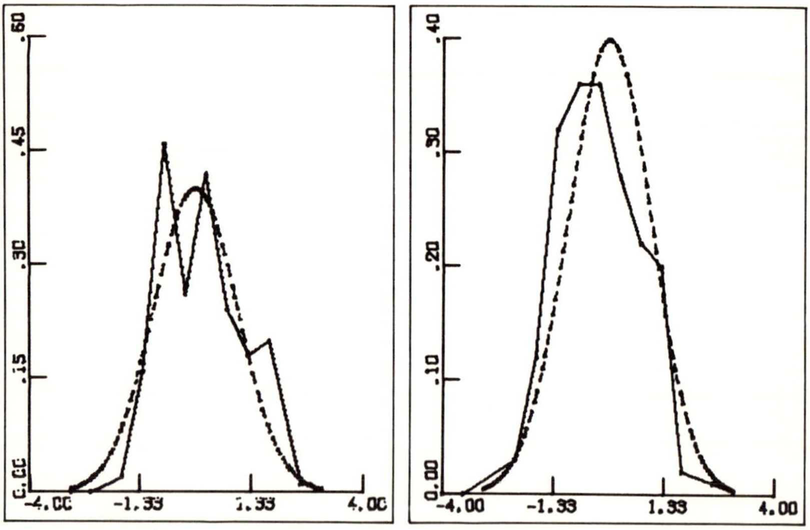

OLS (χ^2 = 17.4, S = 0.31) Likelihood ratio (χ^2 = 10.3)

Fig. 5. Empirical densities of 2 test-procedures (Model 3)

MODEL 4. The Netherlands, 24 pairs of regions, 4 variables.
The regionalization, used in this model is basically the same as in the other models, except that some previous COROP regions are now joined into provinces as follows:

COROP-regions	Province	COROP-regions	Province
1, 2, 3	GRONINGEN (I)	18, 19, 20, 21, 22, 23, 24	NOORD-HOLLAND (VII)
4, 5, 6	FRIESLAND (II)	25, 26, 27, 28, 29, 30	ZUID-HOLLAND (VIII)
7, 8, 9	DRENTHE (III)	31, 32	ZEELAND (IX)
10, 11, 12	OVERIJSSEL (IV)	33, 34, 35, 36	NOORD-BRABANT (X)
13, 14, 15, 16	GELDERLAND (V)	37, 38, 39	LIMBURG (XI)
17	UTRECHT (VI)		

The data are used in a model explaining migration flows to one of the Northern provinces (I, II and III) from any of the other provinces. The regressors are: constant term, ratio of percentages of one-family houses, ratio of sizes of the dependent working population and physical distance of the two provinces' largest towns. Accordingly, we specified a (24×24) W matrix containing weights between flows to or from adjacent provinces and we chose 3 bases, consisting of the following pairs of regions: (IV, I), (IV, II), (IV, III), (IX, II); (IX, I), (IX, II), (IX, III), (XI, II) and (IX, II), (XI, I), (XI, II), (XI, III), where the first number in each pair indicates the starting point of the flow and the second number represents the province of destination. This design yielded the following:

Table 7. Powers of 5 test procedures for Model 4, R = 24, K = 4 (normal error density)

ρ	OLS	BLUS 1	BLUS 2	BLUS 3	RELUS 1	RELUS 2	RELUS 3	LR	NT
0.0	4	7	9	3	4	7	4	0	0
0.1	13	13	12	9	7	8	10	0	5
0.5	76	62	65	65	40	46	60	46	52
0.9	99	92	99	91	86	89	89	99	99
χ^2	14.4	5.9	5.9	5.3	5.2	10.6	9.1	143.8	175.2
S	0.65	0.37	0.37	0.37	0.37	0.37	0.37	–	–

Table 8. Powers of 5 test procedures for Model 4, R = 24, K = 4 (exponential error density)

ρ	OLS	BLUS 1	BLUS 2	BLUS 3	RELUS 1	RELUS 2	RELUS 3	LR	NT
0.0	7	10	9	8	6	9	9	1	1
0.1	8	8	12	10	10	13	10	4	4
0.5	56	40	42	44	31	29	38	34	39
0.9	100	94	98	91	85	87	91	97	98
χ^2	11.2	9.0	6.9	4.5	2.3	6.0	9.4	247.0	223.1

Table 9. Correlation matrices of 4 sets of explanatory variables (determinants parenthesized)

MODEL 1	1.000		MODEL 2	1.000				
(0.987)	-0.147	1.000	(0.409)	-0.147	1.000			
				-0.139	-0.556	1.000		
				0.327	-0.205	0.241	1.000	
				-0.350	0.314	-0.120	-0.132	1.000
MODEL 3	1.000		MODEL 4	1.000				
(0.955)	0.213	1.000	(0.425)	-0.477	1.000			
				-0.269	0.688	1.000		

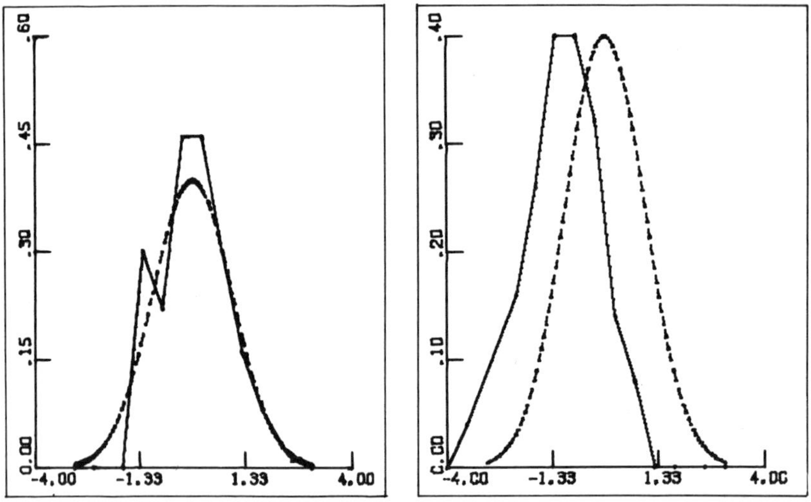

OLS (χ^2 = 14.4, S = 0.65) Likelihood ratio (χ^2 = 143.8)

Fig. 6. Empirical densities of 2 test-procedures (Model 4)

5.8 CONCLUSIONS

In order to judge the performance of the 5 tests on the 4 criteria of section 5.6, we introduce the following scores: −2 means very bad, −1 bad, 0 indifferent, 1 fairly good and 2 stands for very good. We awarded the scores as in Table 10. Because a priori it is not clear which base should be selected when LUS residuals are used, we averaged the scores over the distinct bases. However, the sensitivity to the choice of the base seems to be rather low. The sixth column contains average scores, where weights of 2 and 1 were used for respectively the third criterion (power) and the other criteria. These average scores lead to the ranking of the test-procedures in each model; in case of equal averages, the score on the power determines the rank.

Table 10. Scores of 5 test procedures on 4 criteria

Criterion	(1) Emp. distribution	(2) Sign. level	(3) Power	(4) Robustness	Average	Rank
MODEL 1						
OLS	1	−1	2	1	1.0	1
BLUS	1	−1	1	2	0.8	3
RELUS	2	1	0	2	1.0	2
LR	0	−1	1	−2	−0.2	4
NT	−1	−1	1	−2	−0.4	5
MODEL 2						
OLS	1	2	2	1	1.6	1
BLUS	2	1	0	1	0.8	2
RELUS	1	−1	−1	2	0.0	5
LR	−2	2	1	−2	0.0	4
NT	−2	1	2	−1	0.4	3
MODEL 3						
OLS	−2	−1	2	1	0.4	2
BLUS	−2	0	1	−1	0.6	1
RELUS	1	−1	0	0	0.0	4
LR	1	−1	1	−2	0.0	3
NT	−2	−1	1	−2	−0.6	5
MODEL 4						
OLS	−1	0	2	−1	0.4	1
BLUS	1	−1	1	−1	0.2	2
RELUS	1	0	0	−1	0.0	3
LR	−2	−2	−1	−2	−1.6	5
NT	−2	−2	0	−2	−1.2	4

Table 10 clearly shows that the Moran test with OLS residuals performs best, whereas the Moran test with BLUS residuals may be termed as second-best. Furthermore, it is a striking result that the likelihood ratio test and the naive test do not seem to be appropriate, according to dominant negative scores. Since the regressors were chosen by varying over properities like multicollinearity, autocorrelation and size as much as possible, the results may be interpreted as being fairly general. We conclude with some remarks on further peculiarities.

1. Of all considered Moran statistics, the distribution of the one with OLS residuals has the highest coefficient of skewness in all models. This drawback is compensated for by higher scores on the other criteria. Furthermore, this coefficient is always positive.
2. Except in the models III and IV (OLS), in which autocorrelation and multicollinearity are present in the X-matrix, the distribution of the standardized Moran statistic fits the $N(0, 1)$-distribution well, according to Pearson's χ^2. On the whole the Moran statistic with OLS residuals fitsworse than the tests based on BLUS and RELUS residuals.
3. When the null hypothesis of no spatial autocorrelation is true, the LR test and the naive test produce outcomes which are biased to the left. However, this bias seems to disappear when (R/K) increases (Models 2 and 4 vs. Models 1 and 3).
4. The Moran tests with OLS and BLUS residuals are highly non-robust in model 4. This may be due to the fact that in Model 4 a non-symmetric contiguity matrix is used, whereas $\boldsymbol{W}$ was symmetric in the other models.
5. It cannot be shown that multicollinearity in the matrix of regressors affects the performance of any of the tests significantly. This conclusion was confirmed again by changing the X-matrix of Model 4 such that new matrices with variable degrees of multicollinearity are obtained. Since the results of those experiments also lead to the present conclusions, we do not present them here.

APPENDIX

Bartels and Hordijk (1977) use an alternative definition of the generalized Moran coefficient, in which the LUS-residuals are reduced with their sample-means. The moments of this adjusted Moran coefficient (AMC) are derived below.

We write the $(R-K)$ non-zero elements of (4.1) as

$$\boldsymbol{u}_1 = \boldsymbol{A}_1 \boldsymbol{\epsilon}, \tag{A.1}$$

where $\boldsymbol{A}_1$ is a $(R-K) \times R$ submatrix of $\boldsymbol{A}$. Since the elements of $\boldsymbol{u}_1$ are LUS-residuals, $\boldsymbol{A}_1$ meets the condition

$$A_1A_1' = I_{(R_1)}, \text{ where } R_1 = R - K. \tag{A.2}$$

The idempotent matrix $E = (I_{(R_1)} - (1/R_1)ll')$, in which l is a vector consisting of R_1 unit elements, transforms variables to deviations from their means. The alternative definition of the Moran coefficient now reads as

$$AMC = \frac{\epsilon' A_1' E W_0 E A_1 \epsilon}{\epsilon' A_1' E A_1 \epsilon}, \text{ where } W_0 \text{ is defined as before.} \tag{A.3}$$

Using $tr(W_0) = 0$ and $tr(\mathrm{E}) = R-K-1$, it can easily be derived that

$$\mathrm{E}(AMC) = -\frac{l' W_0 l}{(R-K)(R-K-1)} \tag{A.4}$$

Furthermore

$$\begin{aligned} \mathrm{E}(AMC)^2 = {} & (1/((R-K-1)(R-K+1)))\{tr(W_0W_0') \\ & + tr(W_0^2) - (2/(R-K))(l'W_0W_0'l \\ & + l'W_0^2l) + (3/(R-K)^2)(l'W_0l)^2\} \end{aligned} \tag{A.5}$$

Note that (A.5) is a more complex formula than (4.10).

REFERENCES

Abrahamse, A. P. J. and Koerts, J. (1969), 'A comparison between the power of the Durbin-Watson test and the power of the BLUS test', *Journal of the American Statistical Association*, 64, 938-948.

Bartels, C. P. A. and Bertens, A. M. (1976), 'A factor and regression analysis of regional differences in income-level and -concentration in the Netherlands', *Applied Economics*, 8, 179-192.

Bartels, C. P. A. and Hordijk, L. (1977), 'On the power of the generalized Moran Contiguity Coefficient in testing for spatial autocorrelation among regression disturbances', *Regional Science and Urban Economics*, 7, 83-101.

Cliff, A. D. and Ord, J. K. (1972), 'Testing for spatial autocorrelation among regression residuals', *Geographical Analysis*, 6, 267-284.

Cliff, A. D. and Ord, J. K. (1973), *Spatial autocorrelation*, Pion, London.

Dubbelman, C., Abrahamse, A. P. J. and Koerts, J. (1972), 'A new class of disturbance estimators in the general linear model', *Statistica Neerlandica*, 26, 127-141.

Durbin, J. and Watson, G. S. (1950, 1951, 1971), 'Testing for serial correlation in regression analysis', *Biometrika*, 37, 409-428; 38, 159-178; 58, 1-19.

Fisz, M. (1963), *Probability theory and mathematical statistics*, Wiley, New York.

Hepple, L. W. (1976), 'A Maximum Likelihood model for econometric estimation with spatial data', in I. Masser (ed.), *Theory and practice in regional science*, Pion, London.

Kendall, M. G. and Stuart, A. (1973), *The advanced theory of statistics*, Griffin, London.

Koerts, J. (1967), 'Some further notes on disturbance estimates in regression analysis', *Journal of the American Statistical Association*, 62, 169-183.

Koerts, J. and Abrahamse, A. P. J. (1971), *On the theory and application of the general linear model*, 2nd ed., Rotterdam University Press, Rotterdam.

L'Esperance, W. L. and Taylor, D. (1975), 'The power of four tests of autocorrelation in the linear regression model', *Journal of Econometrics*, 3, 1-21.

Neumann, J. von (1941), 'Distribution of the ratio of the mean square successive difference to the variance', *Annals of Mathematical Statistics*, 12, 367-395.

Ord, J. K. (1975), 'Estimation methods for models of spatial interaction', *Journal of the American Statistical Association*, 70, 120-126.

Phillips, G. D. A. and Harvey, A.C. (1974), 'A simple test for serial correlation in regression analysis', *Journal of the American Statistical Association*, 69, 935-939.

Theil, H. (1965), 'The analysis of disturbances in regression analysis', *Journal of the American Statistical Association*, 60, 1067-1079.

Theil, H. (1971), *Principles of econometrics*, Wiley, New York.

Part 3: Explanatory statistical analysis

6. Multivariate models of dependent spatial data

BERND STREITBERG*

6.1 INTRODUCTION

There exists a sequence of stimulating overviews about (multivariate) data analysis and spatial problems (cf. Berry, 1971; Besag, 1975; Cliff and Ord, 1975; Cliff et al., 1975; Clark et al., 1974; Granger, 1975; Mather and Openshaw, 1974; Unwin and Hepple, 1974). Since I do not feel qualified enough in spatial analysis to add another paper to this admirable list, I have attempted a rather more limited topic: multivariate data analysis in situations where the observations cannot be assumed to be independent.

Imagine you have collected a set of multivariate spatial data, where the observations are most probably dependent, and turn to multivariate statistics for help in analyzing these data. There are basically *three different strategies* open to you:

1. Simply forget about any dependencies and apply the standard textbook procedures. While this is the most usual approach, it is fraught with problems of a quantitative (e.g. tests losing power or not attaining stated significance levels; estimators behaving strangely) and qualitative (e.g. the interpretation of zero order correlations between spatial or temporal series) nature.
2. Try a two-step approach like the one suggested by Cliff and Ord (1972), or Durbin and Watson (1971) in regression analysis. First use the standard procedure (say OLS in regression problems); then test to see if the artificial variables generated by this procedure (residuals, principal components, canoncial

*Freie Universität Berlin, Inst. für Quantitative Ökonomik und Statistik, Corrensplatz 2, 1000 Berlin 33.

variates, cluster assignment vectors) show some form of spatial dependency.

If you can find a suitable test statitic (to my knowledge only tests for regression residuals have been developed so far), this method works well, but only as long as your test does not reject the null hypothesis of independence. Finding 'significant' dependencies indicates that the first step of the procedure may have been invalid – and what then?

3. Assume a particular dependency structure from the beginning and modify or redefine existing methods to take account of this structure.

In the following contribution I introduce a class of very simple dependency structures and try to show how these structures could be used to unify the analysis of dependent data in certain multivariate models: univariate linear models, multivariate linear models, principal components, canonical analysis, discriminant analysis. Since the approaches followed in univariate and multivariate analysis are basically similar, I shall use the univariate case as an example, where the notation is more simple.

Here we have a vector $\boldsymbol{y}$ of observations on N spatial units. We assume that the joint distribution of these observations in hypothetical replications is so civilized that dependencies can be described at least approximately by a covariance matrix $\boldsymbol{\Sigma}$ of order (N, N). Obviously we cannot estimate $N + \binom{N}{2}$ parameters from just N observations, so $\boldsymbol{\Sigma}$ has to be parametrized in some way. A rather general approach to doing this is to postulate that $\boldsymbol{\Sigma}$ has a linear structure, i.e. $\boldsymbol{\Sigma} = \lambda_1 \boldsymbol{K}_1 + \lambda_2 \boldsymbol{K}_2 + \ldots + \lambda_m \boldsymbol{K}_m$, where the (N, N)-matrices $\boldsymbol{K}_s$ are known. The problem is thus reduced to the estimation of m parameters λ_s. While the model looks simple, it is extremely difficult to handle. Consider, for instance, the large number of papers, starting with Hartley and Rao's (1967) breakthrough, that have been written in an effort to make the computation of ML-estimators feasible only for very small data sets.

To achieve further simplification, one has to restrict the set of possible matrices $\boldsymbol{K}_s$ to be used in the parametrization of $\boldsymbol{\Sigma}$. The most simple class of models arises if one assumes that the linear structure is given by the spectral decomposition of Σ, where only the eigenvalues λ_s are unknown. This is the class of models I chose to investigate in this paper.

I would like to remind the reader of the term "spectral decomposition". If you start with a symmetric matrix $\boldsymbol{\Sigma}$, you can compute its eigenvalues λ_s $(s = 1, 2 \ldots m)$ and for each eigenvalue a set of orthogonal eigenvectors $\boldsymbol{x}_{is}$ $(i = 1, 2 \ldots n_s)$ of unit length. If the multiplicity n_s of eigenvalue λ_s is greater than one, the corresponding eigenvectors $\boldsymbol{x}_{is}$ are not uniquely defined (any orthonormal basis will do), but we can construct an idempotent matrix $\boldsymbol{K}_s = \Sigma\, \boldsymbol{x}_{is}\boldsymbol{x}_{is}'$ that characterizes the eigenspace corresponding to λ_s ($\boldsymbol{K}_s$ is the unique projector upon this space). Standard linear algebra then shows that $\boldsymbol{\Sigma}$ can be represented as $\boldsymbol{\Sigma} = \lambda_1 \boldsymbol{K}_1 + \lambda_2 \boldsymbol{K}_2 + \ldots \lambda_m \boldsymbol{K}_m$, where the matrices $\boldsymbol{K}_s$ (the so-called principal idempotents of $\boldsymbol{\Sigma}$) have particularly simple properties (see equation 11). While the covariance structure is simple, it is, to my opinion, not overly restrictive: in section 6.2 a variety of applications is given, including multilevel-structures, spatial moving-average processes, models for transaction tables, etc.

The structure is, of course, not new. It has been investigated for pure random models – where the expectation part is essentially known – in the context of variance-components analysis by Graybill and Huntquist (1961) and for mixed models with $m = 2$ by Balestra (1973). The, rather straightforward, generalization to multivariate models has, to my knowledge, not yet been studied.

Following the introduction of the covariance structure considered in section 6.2, the remaining sections describe univariate and multivariate linear models (section 6.3) and an approach to define principal components (section 6.4). The definition of canonical variates can be undertaken in largely the same way. The concluding section, 6.5, contains some remarks concerning the generalization of cluster and discriminant analysis to dependent data.

6.2 DECOMPOSABLE COVARIANCE STRUCTURES

6.2.1 *Notation*

We are to describe the covariance structure of a random (N, p)-matrix $\boldsymbol{Y}$, usually representing observations of p features on N spatial units. Since, for simplicity of exposition, we do not want to use a coordinate-free approach (as exemplified for instance by Eaton, 1972) we need a convenient notation that allows us to reduce the situation to the familiar one of describing the covariance structure of an Np-vector.

The basic tool is the well-known vec-operator (cf. Nissen, 1968) defining a mapping from the space $\mathbb{R}^{N,p}$ of (N, p)-matrices to the space of Np-vectors $\mathbb{R}^{Np}$. vec $\boldsymbol{Y}$ is a vector containing the elements of $\boldsymbol{Y}$ in lexicographic order of their indices, i.e.

$$\text{vec}\, \boldsymbol{Y} = (y_{11}\ y_{12} \ldots y_{1p}\ y_{21} \ldots y_{2p} \ldots y_{Np})'. \tag{2.1}$$

This mapping preserves *inner products* since (for (N, p)-matrices $\boldsymbol{X}$ and $\boldsymbol{Y}$ and $tr\boldsymbol{A}$ denoting the sum of the diagonal elements of a quadratic matrix $\boldsymbol{A}$):

$$tr\, \boldsymbol{X}'\boldsymbol{Y} = (\text{vec}\, \boldsymbol{X})'(\text{vec}\, \boldsymbol{Y}). \tag{2.2}$$

Using the vec-operator we can describe two linear mappings from $\mathbb{R}^{N,p}$ to $\mathbb{R}^{M,q}$ in a convenient fashion. First consider a *pre- and postmultiplication* of $\boldsymbol{Y}$ by a fixed (M, N)-matrix $\boldsymbol{A}$ and a fixed (p, q) matrix $\boldsymbol{B}'$:

$$\boldsymbol{Y} \longrightarrow \boldsymbol{A}\boldsymbol{Y}\boldsymbol{B}' = \boldsymbol{Z} \tag{2.3}$$

This induces a linear mapping from vec $\boldsymbol{Y}$ to vec $\boldsymbol{Z}$, given by the Kronecker-product $\boldsymbol{A} \otimes \boldsymbol{B}$ of $\boldsymbol{A}$ and $\boldsymbol{B}$:

$$\text{vec}\, \boldsymbol{Y} \longrightarrow (\boldsymbol{A} \otimes \boldsymbol{B})\, \text{vec}\, \boldsymbol{Y} \tag{2.4}$$

where

$$\boldsymbol{A} \otimes \boldsymbol{B} = \begin{bmatrix} a_{11}\boldsymbol{B} & \ldots & a_{1N}\boldsymbol{B} \\ \vdots & & \vdots \\ a_{M1}\boldsymbol{B} & \ldots & a_{MN}\boldsymbol{B} \end{bmatrix} \tag{2.5}$$

Using the resulting equation (cf. Nissen, 1968)

$$\text{vec}\, \boldsymbol{A}\boldsymbol{Y}\boldsymbol{B}' = (\boldsymbol{A} \otimes \boldsymbol{B})\, \text{vec}\, \boldsymbol{Y} \tag{2.6}$$

one easily derives the main properties of Kronecker-products (given for instance by Bellman, 1960).

The second linear mapping we shall consider is *transposition*, i.e. the mapping $\boldsymbol{Y} \longrightarrow \boldsymbol{Y}'$ from $\mathbb{R}^{N,p}$ to $\mathbb{R}^{p,N}$. This induces a corresponding mapping from $\mathbb{R}^{Np}$ to $\mathbb{R}^{Np}$, defined by

$$\boldsymbol{\Phi}\, (\text{vec}\, \boldsymbol{Y}) = (\text{vec}\, \boldsymbol{Y}'). \tag{2.7}$$

$\boldsymbol{\Phi}$ is a permutation matrix of type (Np, Np), since vec Y' contains the same elements as vec Y, in a different order. The following simple properties of $\boldsymbol{\Phi}$ are directly deduced from the defining equation (2.7):

$$\boldsymbol{\Phi}' = \boldsymbol{\Phi}^{-1}, \boldsymbol{\Phi}(A \otimes \boldsymbol{B})\boldsymbol{\Phi} = (\boldsymbol{B} \otimes A) \tag{2.8}$$

$$\boldsymbol{\Phi} = \boldsymbol{\Phi}' \text{ if } N = p \tag{2.9}$$

6.2.2 *Definition*

The class of covariance structures we are going to consider here is given by the following decomposition of var[vec $\boldsymbol{Y}$] – the variance-covariance matrix of vec $\boldsymbol{Y}$ – in a sum of Kronecker-products:

$$\text{var}[\text{vec } \boldsymbol{Y}] = \boldsymbol{K}_1 \otimes \boldsymbol{\Omega}_1 + \boldsymbol{K}_2 \otimes \boldsymbol{\Omega}_2 + \ldots + \boldsymbol{K}_m \otimes \boldsymbol{\Omega}_m \tag{2.10}$$

where the matrices $\boldsymbol{K}_s$ $(s = 1, 2, \ldots, m)$ constitute a set of *known* symmetric matrices of type (N, N), satisfying

$$\boldsymbol{K}_s^2 = \boldsymbol{K}_s, \boldsymbol{K}_s\boldsymbol{K}_t = \boldsymbol{0} \text{ (if } s \neq t), \sum_{s=1}^{m} \boldsymbol{K}_s = \boldsymbol{I} \tag{2.11}$$

and the matrices $\boldsymbol{\Omega}_s$ $(s = 1, 2, \ldots, m)$ are positive-definite (p, p)-matrices usually to be estimated from the data. We shall call this structure a *decomposable covariance structure* with *principal idempotents* $\boldsymbol{K}_s$ and *spectral matrices* $\boldsymbol{\Omega}_s$.

To make this terminology clear, look at the special (univariate) case when Y has just one column. Here we have a random $(N, 1)$-vectory $\boldsymbol{y}$ with covariance matrix

$$\text{var}[\boldsymbol{y}] = \lambda_1 \boldsymbol{K}_1 + \lambda_2 \boldsymbol{K}_2 + \ldots + \lambda_m \boldsymbol{K}_m. \tag{2.12}$$

Assumptions (2.11) imply that $\boldsymbol{K}_s$ is a principal idempotent of var[$\boldsymbol{y}$] with corresponding eigenvalue λ_s. The univariate model is formally analogous to a special class of variance component models that have been investigated for pure 'random models' by Graybill and Huntquist (1961).

6.2.3 *Examples*

A surprisingly large class of dependency structures can be subsumed under the model proposed, possibly after an initial reparametrization. Before embarking on examples with a 'spatial touch', I'd like to introduce the model by means of three well-known structures.

6.2.3.1 *Independence*

In multivariate analysis one usually assumes uncorrelatedness and homoscedasticity of the rows in $\boldsymbol{Y}$. This implies that var[vec $\boldsymbol{Y}$] is given by

$$\text{var}[\text{vec } \boldsymbol{Y}] = \boldsymbol{I} \otimes \boldsymbol{\Omega} = \text{diag}[\boldsymbol{\Omega}, \boldsymbol{\Omega} \ldots \boldsymbol{\Omega}] \tag{2.13}$$

where $\boldsymbol{\Omega}$ is a positive-definite matrix. Note that $\Sigma K_s = I$ guarantees that, regardless of what particular form the matrices K_s take, (2.13) is always a special case of (2.10) if all spectral matrices are equal to Ω.

6.2.3.2 *Heteroscedasticity*

Let the rows of $\boldsymbol{Y}$ be divided into m groups of $r_1, r_2 \ldots r_m$ independent observations with possibly different covariance matrices. Then

$$\text{var}[\text{vec } \boldsymbol{Y}] = \text{diag}[\boldsymbol{\Omega}_1 \ldots \boldsymbol{\Omega}_1, \boldsymbol{\Omega}_2 \ldots \boldsymbol{\Omega}_2, \ldots, \boldsymbol{\Omega}_m \ldots \boldsymbol{\Omega}_m]. \tag{2.14}$$

If we define $\boldsymbol{K}_s$ as a diagonal 0-1-matrix with r_s ones on the main diagonal at positions corresponding to those of the s'th group of observations, (2.14) can be written in the form of (2.10) where the $\boldsymbol{K}_s$ obviously satisfy the assumptions of the decomposable covariance-structure.

In a sense, this situation of heteroscedasticity provides a canonical form into which all decomposable covariance structures can be transformed. To see this, let $\boldsymbol{K}_1, \boldsymbol{K}_2 \ldots \boldsymbol{K}_m$ be a set of matrices satisfying assumption (2.11) and define $r_s := tr\ \boldsymbol{K}_s$. Since $\boldsymbol{K}_s$ is idempotent, there exists a matrix $\boldsymbol{T}_s$ of order (N, r_s), such that $\boldsymbol{T}_s'\boldsymbol{T}_s = \boldsymbol{I}$ and $\boldsymbol{T}_s\boldsymbol{T}_s' = \boldsymbol{K}_s$. The columns of $\boldsymbol{T}_s$ constitute an orthonormal basis of the eigenspace onto which $\boldsymbol{K}_s$ projects. From $\boldsymbol{K}_s\boldsymbol{K}_t = \boldsymbol{0}$ it follows further that $\boldsymbol{T}_s'\boldsymbol{T}_t = \boldsymbol{0}$ if $s \neq t$. The condition $\Sigma \boldsymbol{K}_s = \boldsymbol{I}$ then implies that the matrix $\boldsymbol{T} = [\boldsymbol{T}_1 | \boldsymbol{T}_2 \ldots \boldsymbol{T}_m]$ is an orthogonal (N, N)-matrix. If the covariance structure of $\boldsymbol{Y}$ is given by (2.10), the covariance structure of the transformed data matrix $\boldsymbol{TY}$ is

$$\begin{aligned} \text{var}[\text{vec}\, \boldsymbol{TY}] &= \text{var}[\boldsymbol{T} \otimes \boldsymbol{I})\, \text{vec}\, \boldsymbol{Y}] \\ &= (\boldsymbol{T} \otimes \boldsymbol{I})(\Sigma \boldsymbol{K}_s \otimes \boldsymbol{\Omega}_s)(\boldsymbol{T}' \otimes \boldsymbol{I}) \\ &= \Sigma\, \boldsymbol{TK}_s \boldsymbol{T}' \otimes \boldsymbol{\Omega}_s \end{aligned} \tag{2.15}$$

where

$$\begin{aligned} \boldsymbol{TK}_s\boldsymbol{T}' &= [\boldsymbol{T}_1 \mid \ldots \boldsymbol{T}_m]\, \boldsymbol{T}_s\boldsymbol{T}_s'[\boldsymbol{T}_1 \mid \ldots \boldsymbol{T}_m]' \\ &= \text{diag}\,[\boldsymbol{0}, \boldsymbol{0}, \ldots, \boldsymbol{I}, \boldsymbol{0} \ldots \boldsymbol{0}] \end{aligned} \tag{2.16}$$

is seen to be of the 'heteroscedastic' form. Note that this result entails that the 'effective' number of observations which can be used to estimate the spectral matrix $\boldsymbol{\Omega}_s$ is given by $tr\boldsymbol{K}_s$.

6.2.3.3 *Stationary stochastic processes*

Weak stationary stochastic processes are applied in spatial analysis in two different ways:

1. in the analysis of mixed spatial-temporal data and
2. in the analysis of spatial data with a one-dimensional spatial parameter like in the study of settlement patterns along a highway or in the investigation of transsects through cities.

Segments $\boldsymbol{y} = (Y_1, Y_2 \ldots Y_N)$ of these processes do not have a covariance structure of the type (2.12) for finite N. But there exists an approximate decomposition of var$[\boldsymbol{y}]$ into principal idempotents, which becomes exact for $N \to \infty$ (under mixing assumptions). This approximation allows interpretation of the frequency-analysis of such processes in terms of the decomposable model and motivates the transfer of certain statistical methods used in the analysis of time-series to the general class of dependency structures considered here. To avoid trivial case-distinctions, assume N to be odd, say $N = 2M + 1$. We approximate the distribution of $\boldsymbol{y}$ by that of a circular stochastic process, having autocovariance matrix

$$\begin{aligned} \text{var}[\boldsymbol{y}] &= \gamma_M \boldsymbol{B}^{-M} + \gamma_{M-1} \boldsymbol{B}^{-(M-1)} + \ldots + \gamma_1 \boldsymbol{B}^{-1} + \gamma_0 \boldsymbol{I} \\ &\quad + \gamma_1 \boldsymbol{B} + \ldots + \gamma_M \boldsymbol{B}^M \\ &= \gamma_0 \boldsymbol{I} + \gamma_1 (\boldsymbol{B} + \boldsymbol{B}^{-1}) + \ldots + \gamma_M (\boldsymbol{B}^M + \boldsymbol{B}^{-M}) \end{aligned} \tag{2.17}$$

where the (N, N)-matrix $\boldsymbol{B}$ is the matrix corresponding to the (circular) lag-operator of the process

$$\boldsymbol{B} = (b_{ij}) = \begin{cases} 1 \text{ if } j = 1 + i \pmod N \\ 0 \text{ else} \end{cases} \tag{2.18}$$

Application of $\boldsymbol{B}$ to a vector $(x_1, x_2 \ldots x_N)'$ effects a circular shift of that vector: $\boldsymbol{B}(x_1, x_2 \ldots x_N)' = (x_2, x_3, \ldots x_N, x_1)$. The eigenvalues of $\boldsymbol{B}$ are (cf. Anderson, 1971, p. 281)

$$\omega_\alpha := e^{i 2 \pi (\frac{\alpha}{N})} \text{ for } \alpha = -M, -(M-1), \ldots, (M-1), M.$$

The normalized eigenvector corresponding to ω_α is

$$\boldsymbol{t}_\alpha := N^{-1/2} (\omega_\alpha^{-M}, \omega_\alpha^{-(M-1)}, \ldots \omega_\alpha^{-1}, 1, \omega_\alpha, \ldots \omega_\alpha^{M})', \tag{2.19}$$

where eigenvectors corresponding to different eigenvalues are orthogonal. From the resulting spectral decomposition of the lag operator (where $\boldsymbol{X}^*$ is the conjugate transpose of $\boldsymbol{X}$):

$$\boldsymbol{B} = \sum_{\alpha=-M}^{M} \omega_\alpha \boldsymbol{t}_\alpha \boldsymbol{t}_\alpha^* \tag{2.20}$$

one easily derives the spectral decomposition of $\boldsymbol{B}^s + \boldsymbol{B}^{-s}$:

$$\boldsymbol{B}^s + \boldsymbol{B}^{-s} = \sum_{\alpha=-M}^{M} (\omega_\alpha^s + \omega_\alpha^{-s}) \boldsymbol{t}_\alpha \boldsymbol{t}_\alpha^*. \tag{2.21}$$

Since $\omega_\alpha^s + \omega_\alpha^{-s} = 2 \cos 2\pi \frac{\alpha s}{N}$ this is equal to

$$\boldsymbol{B}^s + \boldsymbol{B}^{-s} = \sum_{\alpha=0}^{M} (2 \cos 2\pi \frac{\alpha s}{N}) \boldsymbol{K}_\alpha$$

where we defined (a vector of all ones is denoted by $\boldsymbol{\iota}$)

$$K_\alpha = \begin{cases} t_0 t_0^* = \dfrac{1}{N}\, \iota\, \iota' \text{ (if } \alpha = 0) \\ t_\alpha t_\alpha^* + t_{-\alpha} t_{-\alpha}^* \text{ (if } \alpha \neq 0). \end{cases} \tag{2.22}$$

K_α is a real, symmetric and idempotent matrix. If we denote the element of K_α in position (j, k) by $k_{jk}^{(\alpha)}$, where j and k run from $-M$ to M, we find

$$k_{jk}^{(\alpha)} = \frac{1}{N}\left(\omega_\alpha^j \omega_\alpha^{-k} + \omega_{-\alpha}^j \omega_{-\alpha}^{-k}\right) = \frac{2}{N} \cos 2\pi \frac{\alpha(j-k)}{N}. \tag{2.23}$$

Since it further can easily be shown, using (2.22), that $K_\alpha K_\beta = 0$ if $\alpha \neq \beta$, and $K_0 + K_1 + \ldots K_M = I$, the set $(K_0, K_1 \ldots K_M)$ satisfies the assumptions of the decomposable model. In terms of the principal idempotents K_α the autocovariance matrix (2.17) of the process can be written

$$\begin{aligned} \text{var}[y] &= \gamma_0 \left(\sum_{\alpha=0}^{M} K_\alpha \right) + 2 \sum_{\tau=1}^{M} \gamma_\tau \left(\sum_{\alpha=0}^{M} \cos 2\pi \frac{\alpha\tau}{N} . K_\alpha \right) \\ &= \lambda_0 K_0 + \lambda_1 K_1 + \ldots + \lambda_M K_M \end{aligned} \tag{2.24}$$

where the linear transformation expressing the eigenvalues λ_α in terms of the autocovariances γ_τ is the fourier-transform of the first $M + 1$ autocovariances $\gamma_0, \gamma_1 \ldots \gamma_M$:

$$\lambda_\alpha = \lambda_0 + 2 \sum_{\tau=1}^{M} \gamma_\tau \cos 2\pi \frac{\alpha\tau}{N}. \tag{2.25}$$

So the eigenvalues of the autovariance matrix of a circular process are approximately equal to the values of the spectral density function at frequencies $\frac{\alpha}{N}$. It is intuitively clear from (2.25) that the approximation by a circular process becomes exact for $N \to \infty$ if the autocovariance function vanishes rapidly enough (i.e. is absolutely summable). A simple proof can be found, for instance, in Fuller (1976).

To obtain an analogous approximation in the multivariate case (i.e. N observations of a vector valued jointly weak stationary process) one would need to generalize definition (2.10) slightly, allowing $\boldsymbol{K}_s$ and $\boldsymbol{\Omega}_s$ to be complex. This generalization is not needed for the spatial applications to follow.

6.2.3.4 *Multilevel structures*

Systems of spatial units are often organized in a hierarchical fashion (i.e. subregions within regions within superregions etc.). Such an organization can be described by an m-tuple $Z_1, Z_2 \ldots Z_m$, where Z_s ($s = 1, \ldots m$) is a partition of the basic set of N spatial units and Z_s is finer than Z_{s+1}. We assume, for convenience, that Z_1 is the finest possible partition, consisting of singletons $\{i\}$, $i = 1, 2 \ldots N$.

Basic to the analysis of such data are operations of *aggregation* (computing subset means) and *disaggregation* (assigning the value of a subset to its members) according to the various 'levels' of the hierarchy defined by the partitions Z_s. If the subsets in Z_s are $M_1, M_2 \ldots M_{N_s}$, we can define two matrices

$$\underset{(N, N_s)}{\boldsymbol{G}_s} = g_{ij}^{(s)} = \begin{cases} 1 \text{ if } i \in M_j \\ 0 \text{ if } i \notin M_j \end{cases} \tag{2.26}$$

$$\underset{(N, N)}{\boldsymbol{\Delta}_s} = \boldsymbol{G}_s (\boldsymbol{G}_s' \boldsymbol{G}_s)^{-1} \boldsymbol{G}_s' \tag{2.27}$$

where $\boldsymbol{G}_s$ effects a disaggregation, $(\boldsymbol{G}_s'\boldsymbol{G}_s)^{-1}\boldsymbol{G}_s'$ aggregates a data vector and $\boldsymbol{\Delta}_s$ assigns aggregated values to the individual spatial units. The above assumption about Z_1 implies $\boldsymbol{\Delta}_1 = \boldsymbol{I}$.

In geometrical terms $\boldsymbol{\Delta}_s$ is seen to project on the space generated by the columns of $\boldsymbol{G}_s$.

Example: $\boldsymbol{Z}_s = (\{1, 2, 5\}, \{3, 4\})$:

$$\boldsymbol{G}_s = \begin{bmatrix} 1 & 0 \\ 1 & 0 \\ 0 & 1 \\ 0 & 1 \\ 1 & 0 \end{bmatrix} \qquad \boldsymbol{\Delta}_s = \begin{bmatrix} 1/3 & 1/3 & 0 & 0 & 1/3 \\ 1/3 & 1/3 & 0 & 0 & 1/3 \\ 0 & 0 & 1/2 & 1/2 & 0 \\ 0 & 0 & 1/2 & 1/2 & 0 \\ 1/3 & 1/3 & 0 & 0 & 1/3 \end{bmatrix}$$

Since data already aggregated to a certain level are not changed by aggregating them again according to the same partition or a finer one, one finds:

1. $\boldsymbol{\Delta}_s^2 = \boldsymbol{\Delta}_s$ with $rg\, \boldsymbol{\Delta}_s = tr\, \boldsymbol{\Delta}_s = N_s$ (2.28)

2. $\boldsymbol{\Delta}_s \boldsymbol{\Delta}_t = \boldsymbol{\Delta}_t \boldsymbol{\Delta}_s = \boldsymbol{\Delta}_t$ (if $s \leq t$) (2.29)

The dependency structure to be introduced can be described informally as an hierarchical moving-average-type model. Let $\boldsymbol{\epsilon}$ be a vector of uncorrelated disturbances with common expectation 0 and variance σ^2. Then we assume that the observed vector $\boldsymbol{y}$ is composed from a systematic part $\mathrm{E}[\boldsymbol{y}]$ and a random part of the form

$$\boldsymbol{u} = (\alpha_1 \boldsymbol{\Delta}_1 + \alpha_2 \boldsymbol{\Delta}_2 + \ldots + \alpha_m \boldsymbol{\Delta}_m)\, \boldsymbol{\epsilon}. \tag{2.30}$$

Note that $\boldsymbol{u}$ is a linear combination of mean disturbances $\boldsymbol{\Delta}_s \boldsymbol{\epsilon}$ averaged according to the various levels of the structure. The coefficients α_s then indicate, for example, the different effects of individual, regional, super-regional influences in the disturbance vector $\boldsymbol{\epsilon}$. Using properties (2.28) and (2.29) one sees that $\mathrm{var}[\boldsymbol{y}]$ then is a linear combination of 'aggregation operators' $\boldsymbol{\Delta}_s$:

$$\mathrm{var}[\boldsymbol{y}] = \gamma_1 \boldsymbol{\Delta}_1 + \gamma_2 \boldsymbol{\Delta}_2 + \ldots + \gamma_m \boldsymbol{\Delta}_m \tag{2.31}$$

A simple reparameterization allows this to be written in the form (2.12) of the reducible model. Defining

$$\boldsymbol{K}_s := \begin{cases} \boldsymbol{\Delta}_s - \boldsymbol{\Delta}_{s+1} & \text{if } s = 1, 2 \ldots m-1 \\ \boldsymbol{\Delta}_m & \text{if } s = m, \end{cases} \tag{2.32}$$

the properties (2.29), (2.30) show that the matrices $\boldsymbol{K}_1, \boldsymbol{K}_2 \ldots \boldsymbol{K}_m$ indeed satisfy (2.11).

If $\mathrm{var}[\boldsymbol{y}]$ is expressed in terms of these principal idempotents, it is easily found that

$$\mathrm{var}[\boldsymbol{y}] = \lambda_1 \boldsymbol{K}_1 + \lambda_2 \boldsymbol{K}_2 + \ldots + \lambda_m \boldsymbol{K}_m \tag{2.33}$$

where $\lambda_s = \gamma_1 + \gamma_2 + \ldots + \gamma_s$. The corresponding multivariate structure

$$\mathrm{var}[\mathrm{vec}\,\boldsymbol{y}] = \boldsymbol{K}_1 \otimes \boldsymbol{\Omega}_1 + \boldsymbol{K}_2 \otimes \boldsymbol{\Omega}_2 + \ldots + \boldsymbol{K}_m \otimes \boldsymbol{\Omega}_m \tag{2.34}$$

can be derived in a completely analogous fashion. If we picture a hierarchical data situation as a tree, where nodes at a certain level correspond to the subsets of a partition, one can identify the principal idempotents $\boldsymbol{K}_s$ with the branches connecting adjacent levels of this tree.

6.2.3.5 *Transaction tables*

One important task of regional economics or geography is to analyze flows or transactions occurring between spatial locations like commodity streams in regional input-output systems, journeys, migrations, interregional marriages, etc. The data appear in the form of one or more transaction tables $\boldsymbol{F} = (f_{ij})$, where f_{ij} is the flow from location i to location j. We assume $\boldsymbol{F}$ to be quadratic of type (n, n), but the case of rectangular tables (i.e. flows from location i in country A to location j in country B) can be handled in a similar fashion.

In modelling the expectation part of transaction tables, often gravity analogies are used. An alternative could be the multivariate analysis of a set of such tables, possibly including distance-tables. For a different approach compare the linear decomposition models of Kuhbier (1976), that allow one to handle extraneous information about the spatial locations in a very convenient fashion. We shall focus here, of course, on the covariance-part of such models. It is obvious that the usual assumption of independence between all different elements (f_{ij}) is often not justified.

A relatively simple alternative is generated, if we assume that dependencies between flows f_{ij} and f_{kl} are caused by dependencies between "row"-locations i, k on the one hand and "column"-locations j, l on the other. Let $(\boldsymbol{R}_1, \boldsymbol{R}_2 \ldots \boldsymbol{R}_m)$ be a set of principal idempotents describing dependencies between row locations $-\boldsymbol{R}_s$ is of type (n, n) $-$ and $(\boldsymbol{C}_1, \boldsymbol{C}_2 \ldots \boldsymbol{C}_k)$ a corresponding set for column locations. A decomposable covariance structure of $\boldsymbol{f} = \mathrm{vec}[\boldsymbol{F}]$ is then given by all possible Kronecker-products of these idempotents, i.e.

$$\mathrm{var}[\boldsymbol{f}] = \sum_{s=1}^{m} \sum_{t=1}^{k} \lambda_{st} \boldsymbol{R}_s \otimes \boldsymbol{C}_t. \tag{2.35}$$

If we wished to analyze the joint structure of a set $\boldsymbol{F}_1, \boldsymbol{F}_2 \ldots \boldsymbol{F}_p$ of transaction tables (say: interregional input-output flows of p different commodities), the corresponding multivariate model could be used:

$$\operatorname{var}[\operatorname{vec} \boldsymbol{Y}] = \sum_{s=1}^{m} \sum_{t=1}^{k} \boldsymbol{R}_s \otimes \boldsymbol{C}_t \otimes \boldsymbol{\Omega}_{st} \qquad (2.36)$$

where the w 'th column of $\boldsymbol{Y}$ contains the n^2-vector vec $\boldsymbol{F}_w$.

Example. Let a (2, 2)–table:

$$\boldsymbol{F} = \begin{bmatrix} f_{11} & f_{12} \\ f_{21} & f_{22} \end{bmatrix} \qquad (2.37)$$

be given, i.e. $\boldsymbol{f} = (f_{11}\ f_{12}\ f_{21}\ f_{22})'$.

Assuming constant row-wise and constant column-wise dependencies, a possible covariance structure between the elements of $\boldsymbol{f}$ could be

$$\operatorname{var}[\boldsymbol{f}] = \begin{bmatrix} \gamma_1+\gamma_2+\gamma_3+\gamma_4 & \gamma_2+\gamma_4 & \gamma_3+\gamma_4 & \gamma_4 \\ \gamma_2+\gamma_4 & \gamma_1+\gamma_2+\gamma_3+\gamma_4 & \gamma_4 & \gamma_3+\gamma_4 \\ \gamma_3+\gamma_4 & \gamma_4 & \gamma_1+\gamma_2+\gamma_3+\gamma_4 & \gamma_2+\gamma_4 \\ \gamma_4 & \gamma_3+\gamma_4 & \gamma_2+\gamma_4 & \gamma_1+\gamma_2+\gamma_3+\gamma_4 \end{bmatrix} \qquad (2.38)$$

In matrix form (where $\boldsymbol{J}$ is a (2, 2)-matrix of ones):

$$\operatorname{var}[\boldsymbol{f}] = \gamma_1 \boldsymbol{I} \otimes \boldsymbol{I} + \gamma_2 \boldsymbol{I} \otimes \boldsymbol{J} + \gamma_3 \boldsymbol{J} \otimes \boldsymbol{I} + \gamma_4 \boldsymbol{J} \otimes \boldsymbol{J}. \qquad (2.39)$$

We see that γ_1 can be interpreted as a variance parameter and γ_2, γ_3, γ_4 as column, row, and overall covariance respectively. A reparametrization shows this to be of the type (2.36):

$$\operatorname{var}[\boldsymbol{f}] = \gamma_1 \bar{\boldsymbol{M}} \otimes \bar{\boldsymbol{M}} + (\gamma_1 + \gamma_2\, n) \bar{\boldsymbol{M}} \otimes \boldsymbol{M}$$

$$+ (\gamma_1 + \gamma_3\, n) \boldsymbol{M} \otimes \bar{\boldsymbol{M}} \qquad (2.40)$$

$$+ (\gamma_1 + \gamma_2 n + \gamma_3 n + \gamma_4 n^2) \boldsymbol{M} \otimes \boldsymbol{M}$$

where $\boldsymbol{M} = \frac{1}{n} \boldsymbol{J}; \bar{\boldsymbol{M}} = \boldsymbol{I} - \boldsymbol{M}$.

The matrices $(\boldsymbol{M}, \bar{\boldsymbol{M}})$ define a multi-level structure of the simplest possible kind, consisting of just two partitions Z_1, Z_2 where Z_1 is the finest possible partition and Z_2 is the coarsest one (i.e. lumping together all locations in one set). Of course, the example is purely illustrational and nobody is advised actually to use such a model to analyze a (2,2)-table.

6.2.3.6 *Transaction tables: reciprocal dependencies*

A more interesting model of dependencies in transaction tables emerges if 'reciprocal' elements in the table (f_{ij} and f_{ji}) are considered. Often a strong dependency between a flow from location i to j and the corresponding flow from location j back to i should be assumed out of substantive reasons (consider for example the case of journeys between different countries). One possible model resulting from this idea, with particularly simple properties, will be illustrated by using the (2, 2)-table of 6.2.3.5.

Example. The covariance-structure is given by:

$$\operatorname{var}[\boldsymbol{f}] = \left[\begin{array}{cc|cc} \alpha+\beta & 0 & 0 & 0 \\ 0 & \alpha & \beta & 0 \\ \hline 0 & \beta & \alpha & 0 \\ 0 & 0 & 0 & \alpha+\beta \end{array}\right] \tag{2.41}$$

Note the appearance of the covariance parameter β between 'reciprocal' elements (f_{12}, f_{21}), (f_{11}, f_{11}), (f_{22}, f_{22}). If we use the transposition operator $\boldsymbol{\Phi}$ defined in equation (2.8), the model can be written in matrix form

$$\operatorname{var}[\boldsymbol{f}] = \alpha \boldsymbol{I} + \beta \boldsymbol{\Phi}. \tag{2.42}$$

Noting that the matrices $\frac{1}{2}(\boldsymbol{I} + \boldsymbol{\Phi})$ and $\frac{1}{2}(\boldsymbol{I} - \boldsymbol{\Phi})$ are idempotent and their product is $\boldsymbol{0}$, a reparametrization transforms this into a reducible structure:

$$\text{var}[f] = (\alpha+\beta)\frac{1}{2}(I+\Phi) + (\alpha-\beta)\frac{1}{2}(I-\Phi). \tag{2.43}$$

As in 6.2.3.4, where we interpreted the principal idempotents using a corresponding decomposition of the data, we see that $\frac{1}{2}(I+\Phi)$ and $\frac{1}{2}(I-\Phi)$ effect a decomposition of F in two parts:

$$F = \frac{1}{2}(F+F') + \frac{1}{2}(F-F') \tag{2.44}$$

a symmetric one, where reciprocal elements are identified and a skew-symmetric one, where reciprocal flows are contrasted. Since $tr\, I = n^2$ and $tr\, \Phi = n$, the number of observations available to estimate $(\alpha+\beta)$ and $(\alpha-\beta)$ are $\frac{1}{2}n(n+1)$ and $\frac{1}{2}n(n-1)$ respectively.

The covariance structure considered here and the one considered in section 6.2.3.5 can be combined. I shall only give an example where the idempotents used are $(M, \bar{M})$ for rows and for columns:

$$\begin{aligned}\text{var}[f] = {} & \lambda_1 \frac{1}{2}(I+\Phi)(\bar{M}\otimes\bar{M}) + \lambda_2 \frac{1}{2}(I-\Phi)(\bar{M}\otimes\bar{M}) \\ & + \lambda_3 \frac{1}{2}(I+\Phi)(M\otimes\bar{M} + \bar{M}\otimes M) \\ & + \lambda_4 \frac{1}{2}(I-\Phi)(M\otimes\bar{M} + \bar{M}\otimes M) + \lambda_5\, M\otimes M.\end{aligned}$$

There's quite some room for variety here, if more refined multiple-level-structures are used instead of $(M, \bar{M})$.

6.2.3.7 *Dependencies within regions*

The following example is a generalization of a two-level-covariance structure.

Consider partition (Z_1, Z_2), defining a hierarchical situation with two levels (i.e. a grouping of N spatial units in m regions). We assume the following covariance matrix of observations belonging to the r'th region:

$$\boldsymbol{\Sigma}_r = \begin{bmatrix} \gamma_0 + \gamma_r & \gamma_r & \dots & \gamma_r \\ \gamma_r & \gamma_0 + \gamma_r & \dots & \gamma_r \\ \vdots & \vdots & \ddots & \\ \gamma_r & \gamma_r & & \gamma_0 + \gamma_r \end{bmatrix} \tag{2.45}$$

Here γ_r is a region-specific covariance-parameter and γ_0 determines the variance of individual observations. Now let var$[\boldsymbol{y}]$ be the direct sum of these intragroup-covariance matrices, i.e.

$$\text{var}[\boldsymbol{y}] = \text{diag}[\Sigma_1, \Sigma_2 \dots \Sigma_m]. \tag{2.46}$$

Again this can be reparameterized to be of the decomposable type. If Z_2 consists of m subsets $M_1, M_2 \dots M_m$ with $N_1, N_2 \dots N_m$ elements, we define (N, N)-matrices $\boldsymbol{M}_r$ by

$$\boldsymbol{M}_r = (m_{ij}^{(r)}) = \begin{cases} \dfrac{1}{N_r} & \text{if } i \in M_r \text{ and } j \in M_r \\ 0 & \text{else} \end{cases} \tag{2.47}$$

and a matrix

$$\boldsymbol{I} - \boldsymbol{\Delta} = \boldsymbol{I} - (\boldsymbol{M}_1 + \boldsymbol{M}_2 + \dots \boldsymbol{M}_m). \tag{2.48}$$

Then one easily finds that var$[\boldsymbol{y}]$ can be written in the form of (2.12):

$$\text{var}[\boldsymbol{y}] = \lambda_0 (\boldsymbol{I} - \boldsymbol{\Delta}) + \lambda_1 \boldsymbol{M}_1 + \dots + \lambda_m \boldsymbol{M}_m \tag{2.49}$$

where $\lambda_0 = \gamma_0$, $\lambda_r = N_r \gamma_r + \gamma_0$.

As it stands, this model is not very useful for estimation purposes, since the effective number of observations available to estimate λ_r is only equal to $tr\boldsymbol{M}_r = 1$. A way out is to impose further restrictions on the eigenvalues. In general, linear restrictions can be given either in *parametric form*:

$$\boldsymbol{\lambda} = \boldsymbol{C} \alpha \tag{2.50}$$

where $\boldsymbol{\lambda} = (\lambda_1, \lambda_2 \dots \lambda_m)'$, and $\alpha = (\alpha_1, \dots a_s)'$ is a new set of parameters, or in *constraint form*:

$$D\lambda = 0 . \tag{2.51}$$

Example. The two-level structure results from (2.49) if one constrains all $\lambda_1 \ldots \lambda_m$ to be equal. Here (note there's only one new parameter)

$$C = \begin{bmatrix} 1 \\ 1 \\ \cdot \\ \cdot \\ \cdot \\ 1 \end{bmatrix} \qquad D = \begin{bmatrix} 1 & -1 & & & \\ & 1 & -1 & & \\ & & 1 & -1 & \\ & & & \ddots & \\ & & & & -1 \end{bmatrix} \tag{2.52}$$

By allowing the user to impose linear constrains on the parameters, the model can be adapted to a larger variety of substantive assumptions.

6.2.3.8 *Spatial moving-average processes*
Spatial moving-average models (cf. Cliff and Ord, 1975)

$$y = (I + \alpha_1 W + \alpha_2 W^2 + \ldots + \alpha_r W^r)\, \epsilon \tag{2.53}$$

where $E[\epsilon] = 0$, $\mathrm{var}[\epsilon] = \sigma^2 I$ and W is a known weight-matrix (here assumed to be symmetric), imply a covariance structure of the following kind:

$$\mathrm{var}[y] = \gamma_0 I + \gamma_1 W + \gamma_2 W^2 + \ldots + \gamma_s W^s . \tag{2.54}$$

These covariance-structures may be of interest in their own right being analogous to s-dependent stationary stochastic processes.

We transform (2.54) to a decomposable form by using the spectral representation of W, found via an eigenanalysis of the weight-matrix W giving eigenvalues p_r and principal idempotents K_r:

$$W = p_1 K_1 + p_2 K_2 + \ldots + p_m K_m . \tag{2.55}$$

Here $(K_1, K_2 \ldots K_m)$ is a set of matrices, fulfilling assumption (2.11). Then

$$\mathrm{var}[y] = \lambda_1 K_1 + \lambda_2 K_2 + \ldots + \lambda_m K_m \tag{2.56}$$

where

$$\lambda_r = \gamma_0 + \gamma_1 p_r + \gamma_2 p_r^2 + \ldots + \gamma_s p_r^s. \tag{2.57}$$

Note that in general the new parameters λ_r are not free to vary over R_+ but are restricted by the relations (2.57). Since W is known, its eigenvalues p_r are known and we can write these restrictions e.g. in parametric form with the help of a known matrix $\boldsymbol{C} = (c_{ij}) = p_i^j$ $(i = 1, 2, \ldots, m; j = 1, 2, \ldots, s)$ as $\boldsymbol{\lambda} = \boldsymbol{C\gamma}$.

Instead of (2.55) where in general N principal idempotents are needed (if all eigenvalues of $\boldsymbol{W}$ are different), an *approximate* spectral decomposition of $\boldsymbol{W}$ could be used as outlined in the following section.

6.2.3.9 *An empirical method to generate decomposable structures*

In analyzing purely spatial data (the situation is much better in mixed time-series/spatial data, cf. Granger, 1975), our knowledge about the actual spatial dependency structure will often be extremely vague. We need models where one can choose from a large class of alternative covariance assumptions to be tentatively entertained, so that one is able to check how strongly substantive results change when the dependency model is varied.

The following is a suggestion – admittedly entirely empirical – how one could use vague a priori knowledge to develop covariance structures of the decomposable type.

Given N sites, one often starts with a (symmetric) weight matrix $\boldsymbol{W}$, representing some a priori knowledge about dependencies in the spatial structure. If one could decompose $\boldsymbol{W}$ as a linear combination of m (where m is hopefully small) idempotent and pairwise orthogonal matrices $\boldsymbol{K}_s$ (hopefully representing certain basic features of the spatial system) a possible model for var[$\boldsymbol{y}$] would be

$$\text{var}[\boldsymbol{y}] = \sum_{s=1}^{m} \lambda_s \boldsymbol{K}_s$$

where λ_s is a set of parameters to be estimated from the data.

There's a natural way in which such a decomposition arises in the eigenanalysis of the weight matrix. In general, all eigenvalues p_s of $\boldsymbol{W}$ will be different and the decomposition in principal idempotents will have N terms, which is not very useful for estimation purposes.

Since eigenvectors corresponding to close eigenvalues are not well determined, being subject to large variation if $\boldsymbol{W}$ is changed just a bit, it is tempting to partition the set of eigenvaleus in m groups and approximate $\boldsymbol{W}$ by

$$\bar{\boldsymbol{W}} = \sum_{i=1}^{m} \bar{p}_i \boldsymbol{K}_i$$

where $\bar{p}_i$ is an average of the eigenvalues in the i'th group and $\boldsymbol{K}_i = \boldsymbol{T}_i \boldsymbol{T}_i'$ with $\boldsymbol{T}_i$ having as columns the eigenvectors corresponding to these eigenvalues.

The following cluster analysis procedure may be followed, in case no substantive information is available for grouping the eigenvalues. If one measures the distance between $\boldsymbol{W}$ and $\bar{\boldsymbol{W}}$ by $[tr(\boldsymbol{W}-\bar{\boldsymbol{W}})^2]^{1/2}$, a minimal-distance partition, given m, can be found exactly by an adaptation of Fisher's dynamic programming procedure (Fisher, 1958). This 'optimal' partition will consist of m groups of *adjacent* eigenvalues. An idea of the best m can be obtained by plotting the minimal-distances against m and noting where spectacular reductions take place.

Experience with this procedure in principal components analysis suggests that the resulting idempotents are very often much easier to interpret than in the original eigenvectors – it is helpful that all elements of $\boldsymbol{K}_i$ lie in the interval $[-1, 1]$. If the idempotents can be identified with distinct features of the spatial system, the corresponding parameters will be directly interpretable. In any case, an estimator for var$[\boldsymbol{y}]$ is obtained.

6.3 LINEAR MODELS

6.3.1 *Introduction*

Linear models (regression analysis, analysis of variance) constitute one major application of statistical methods in spatial analysis. One way to deal with the dependencies of the observations (or equivalently, the disturbances) in these models is to estimate the parameters in the usual way by least-squares and then test if there is some form of spatial dependency in the residuals, apart from the dependencies introduced by the estimation procedure. This is the approach followed for instance by Cliff and Ord (1972). Since OLS-residuals are always dependent, alternatives like BLUS-residuals have been investigated (cf. Bartels, 1976, and references given there).

An alternative procedure would be to incorporate dependency assumptions directly in the model. This is the approach followed here. The main problem of this approach of course is the commitment to a particular dependency structure. By reestimating the model with different dependency assumptions the effects of this commitment can and should be studied.

Let $\boldsymbol{y}$ be a random $(n, 1)$-vector with expectation $\mathrm{E}[\boldsymbol{y}] = \boldsymbol{X\beta}$ and $\mathrm{var}[\boldsymbol{y}] = \boldsymbol{\Sigma}$. For short, this linear model is written $(\boldsymbol{y}, \boldsymbol{X\beta}, \boldsymbol{\Sigma})$. Specifically, *two* types of models will be considered which differ in assumptions about the expectation part ((B) results from (A) by specializing the $\boldsymbol{X}$-matrix):

$$\begin{aligned} &(A) \quad (\boldsymbol{y}, \boldsymbol{X\beta}, \Sigma \lambda_s \boldsymbol{K}_s) \\ &(B) \quad (\boldsymbol{y}, \Sigma \boldsymbol{X}_s \boldsymbol{\beta}_s, \Sigma \lambda_s \boldsymbol{K}_s) \end{aligned} \qquad (3.1)$$

where in type (A) models, $\boldsymbol{X}$ is assumed to be of full column rank k and in type (B) models, $\boldsymbol{X}_s = \boldsymbol{K}_s\boldsymbol{X}$, again $\boldsymbol{X}$ being of full column rank k.

To make the distinction between these types clear, consider the case of a stationary process. While (A) assumes that effects of the 'independent variables' $\boldsymbol{X}$ do not vary with frequency, a model of type (B) allows explicitly for differential effects in different frequency bands (β_s is the vector valued transfer function at frequency s). In multi-level structures a type (B) model assumes that the same independent variables behave differently according to their aggregation level. Models of type (B) are much simpler to handle than models of type (A).

If we assume multivariate normality of $\boldsymbol{y}$ both models are seen to be members of the exponential class of distributions. In type (A) models the minimal sufficient statistic is (in general) the $m \cdot (k+1)$-dimensional statistic $(\boldsymbol{y}'\boldsymbol{K}_1\boldsymbol{y}, \ldots, \boldsymbol{y}'\boldsymbol{K}_m\boldsymbol{y}, \boldsymbol{X}'\boldsymbol{K}_1\boldsymbol{y}, \ldots \boldsymbol{X}'\boldsymbol{K}_m\boldsymbol{y})$. It is not difficult to show (e.g. Streitberg, 1977) that a necessary and sufficient condition for minimal sufficiency of the $(m + k)$-dimensional statistic $(\boldsymbol{y}'\boldsymbol{K}_1\boldsymbol{y}, \ldots, \boldsymbol{y}'\boldsymbol{K}_m\boldsymbol{y}, \boldsymbol{X}'\boldsymbol{y})$ is the invariance of the column space of $\boldsymbol{X}$ under all $\boldsymbol{K}_s$, i.e. the existence of matrices $\boldsymbol{L}_1, \boldsymbol{L}_2 \ldots \boldsymbol{L}_m$ such that $\boldsymbol{K}_s\boldsymbol{X} = \boldsymbol{X}\boldsymbol{L}_s$. The said statistic is then also completely sufficient, using Theorem 1 of Lehmann (1959, p. 132).

In model (B) the 'design-matrix' is $[\boldsymbol{X}_1 \mid \boldsymbol{X}_2 \ldots \boldsymbol{X}_m]$ and one finds

$$K_s[X_1 \mid X_2 \mid \ldots X_s \ldots \mid X_m] = [0 \mid 0 \mid \ldots X_s \ldots \mid 0]$$
$$= [X_1 \mid X_2 \mid \ldots X_s \ldots \mid X_m]$$
$$\text{diag}[0, 0, \ldots, I, \ldots, 0] \quad (3.2)$$

so that the conditions for complete sufficiency of a statistic equal in dimension to the parameter space are satisfied and important simplifications of estimation procedures follow.

6.3.2 *Estimation in unrestricted models*

6.3.2.1 *Overview*

Under the model $(y, X\beta, \Sigma\, \lambda_s K_s)$ the observed y is a random point of $\mathbb{R}^N$ which we can regard as a superposition of $(m + 1)$ vectors

$$y = \mu + u_1 + u_2 + \ldots + u_m \quad (3.3)$$

where $\mu (= X\beta)$ is a non-random vector known to lie in the column space $R(X)$ of X, and $u_s = K_s(y-\mu)$ is an unobservable vector of disturbances corresponding to 'level' s of the dependency structure. Our goal is to recover this decomposition as far as possible.

If all λ_s were known, the standard Gauss-Markov theory would immediately lead to a decomposition of y into μ and $u_1 + u_2 + \ldots + u_m$ via a projection of y onto the regression space $R(X)$ and its orthogonal complement with the projectors $P_\Sigma := X(X'\Sigma^{-1}X)^{-1}X'\Sigma^{-1}$ and $Q_\Sigma := I - P_\Sigma$. On the other hand, if μ were known, the components u_s would be found by projecting $(y-\mu)$ onto the subspace corresponding to K_s. The standard variance-components-theory shows (cf. Graybill and Huntquist, 1961) that the quadratic forms

$$\hat{\lambda}_s = (tr\,K_s)^{-1}(y-\mu)'K_s(y-\mu) = (tr\,K_s)^{-1}\hat{u}_s'\hat{u}_s \quad (3.4)$$

where $\hat{u}_s = K_s(y-\mu)$, are unbiased complete sufficient statistics and thus the best unbiased estimators of λ_s (under normality assumptions).

These remarks lead us to consider two alternative strategies. The first one starts with some preliminary estimator $\hat{\mu}$ of μ (for instance the OLS-estimator $Py = X(X'X)^{-1}X'y$ or a GLS-estimator if one has some a priori knowledge about the ratios of eigenvalues λ_s). The *residuals* $\boldsymbol{e} = \boldsymbol{y} - \hat{\boldsymbol{\mu}}$ are then decomposed via $\boldsymbol{e}_s = \boldsymbol{K}_s\boldsymbol{e}$ and quadratic functions of $\boldsymbol{e}$ like (3.4) used to estimate λ_s. The procedure can be iterated, i.e. $\hat{\lambda}_s$ is used to obtain a new projector $\boldsymbol{P}_{\hat{\Sigma}}$ and new decompositions $\boldsymbol{K}_s$ $(\boldsymbol{I}-\boldsymbol{P}_{\hat{\Sigma}})\boldsymbol{y}$. It is remarkable that this iteration procedure yields ML-estimators of $\boldsymbol{\beta}$ and λ_s.

The second strategy projects the original observations onto the eigenspace of $\boldsymbol{K}_s$, i.e. starts with decompositions $\boldsymbol{K}_s\boldsymbol{y}$. These projections follow a *partial-model* $(\boldsymbol{K}_s\boldsymbol{y}, \boldsymbol{K}_s\boldsymbol{X}\boldsymbol{\beta}, \lambda_s\boldsymbol{K}_s)$, having essentially the form of the standard linear model. The standard least-squares theory leads to unbiased estimators of λ_s, possessing useful distributional properties. These estimators can be used to derive two-phase estimators of $\boldsymbol{\beta}$.

6.3.2.2 *Assumptions for asymptotics*

Some of the more complicated estimators necessitate asymptotic considerations, since finite sample properties are intractable. We have to define sequences of linear models with a growing number of observations and then study the behaviour of our estimators in such sequences. In models with dependent data, the problem obviously arises that the definition of these sequences is in no way unique, since addition of observations to a given 'sample' can change the covariance matrix in many conceivable ways. Consider for example a two level structure with N spatial units and N_1 regions. Here the addition of observations can have quite different effects:

1. group-sizes grow, while N_1 stays constant (e.g. a finer spatial grid is used within the regions);
2. group-sizes stay constant and N_1 grows (e.g. further regions are considered);
3. both group-sizes and number of regions go up;
4. the structure itself changes into a three-level structure (e.g. by going from the analysis of one country to that of a whole system of countries).

We shall exclude asymptotic situations of the type exemplified under (4), i.e. the number of parameters growing over all bounds when $N \to \infty$. Assumptions of this type would be needed in time-series analysis, but in none of the spatial applications considered.

Since the effective number of observations available to estimate λ_s is $r_s = tr\,\boldsymbol{K}_s$, we shall assume that observations are added in such a way that all r_s grow beyond limits with N and no particular level of the structure eventually dominates the system or eventually vanishes, i.e.

$$\lim_{N \to \infty} r_s(N) = \infty \quad \text{and} \quad \lim_{N \to \infty} \frac{r_s(N)}{N} = k_s \in (0, 1) \tag{3.5}$$

These asymptotic assumptions and assumptions about the expectation part (effectively uniform boundedness of the sequence of regressor matrices $\boldsymbol{X}$) and about cumulants of $\boldsymbol{y}$ are delineated more carefully in Streitberg (1977).

6.3.2.3 *Estimators, using strategy 1*

Let the positive definite matrix $\boldsymbol{\Sigma}_0 = \lambda_1^{(0)} \boldsymbol{K}_1 + \ldots + \lambda_m^{(0)} \boldsymbol{K}_m$ represent an a priori assumption about the covariance structure of y. The GLS-estimator of $\boldsymbol{\beta}$ then is

$$\dot{\boldsymbol{\beta}} = (\boldsymbol{X}'\boldsymbol{\Sigma}_0^{-1}\boldsymbol{X})^{-1}\boldsymbol{X}'\boldsymbol{\Sigma}_0^{-1}\boldsymbol{y} \tag{3.6}$$

and the GLS-residuals are defined by $\dot{\boldsymbol{e}} = \dot{\boldsymbol{Q}}\boldsymbol{y}$ where

$$\dot{\boldsymbol{Q}} = \boldsymbol{I} - \boldsymbol{X}(\boldsymbol{X}'\boldsymbol{\Sigma}_0^{-1}\boldsymbol{X})^{-1}\boldsymbol{X}'\boldsymbol{\Sigma}_0^{-1} \tag{3.7}$$

An intuitively reasonable estimator of λ_s using these residuals then is the observable analogue of (3.4):

$$\dot{\lambda}_s = (r_s)^{-1}\dot{\boldsymbol{e}}'\boldsymbol{K}_s\dot{\boldsymbol{e}} = (r_s)^{-1}\boldsymbol{y}'\dot{\boldsymbol{Q}}'\boldsymbol{K}_s\dot{\boldsymbol{Q}}\boldsymbol{y}. \tag{3.8}$$

The following properties are easily derived (cf. Streitberg, 1977):

1. $\dot{\lambda}_s$ is a *non-negative* quadratic form;

2. $\dot{\lambda}_s$ is *translation-invariant*, i.e. if the data $\boldsymbol{y}$ are transformed to $\boldsymbol{y} + \boldsymbol{X}\boldsymbol{a}$, $\dot{\lambda}_s$ is not changed;

3. $\dot{\lambda}_s$ is *biased* in general. If we define $\dot{\boldsymbol{\lambda}} = (\dot{\lambda}_1, \ldots, \dot{\lambda}_m)'$; $\boldsymbol{\lambda} = (\lambda_1, \ldots, \lambda_m)'$; $\boldsymbol{V} = (v_{st}) = (r_s)^{-1}(tr\, \dot{\boldsymbol{Q}}'\boldsymbol{K}_s\dot{\boldsymbol{Q}}\boldsymbol{K}_t)$, one finds

$$\mathrm{E}[\dot{\boldsymbol{\lambda}}] = \boldsymbol{V}\boldsymbol{\lambda}\ ; \tag{3.9}$$

4. The covariance between $\dot{\lambda}_s$ and $\dot{\lambda}_t$ is given by

$$\operatorname{cov}[\dot{\lambda}_s, \dot{\lambda}_t] = \frac{2}{r_s r_t} \boldsymbol{\lambda}' V^{(s,t)} \boldsymbol{\lambda} \tag{3.10}$$

where

$$V^{(s,t)} = (v_{ij}^{(s,t)}) = (tr\, \dot{Q}'K_s\dot{Q}K_i\dot{Q}'K_t\dot{Q}K_j); \tag{3.11}$$

5. $\dot{\lambda}_s$ and $\dot{\boldsymbol{\beta}}$ are uncorrelated, but in general not independent.

Note that in (4) and (5) multivariate normality of $\boldsymbol{y}$ is needed. If the asymptotic assumptions are satisfied, one can show (cf. Streitberg, 1977, p. 77) that

6. $\dot{\lambda}_s$ is asymptotically unbiased and consistent;

7. $\dot{\lambda}_s$ and $\dot{\lambda}_t$ $(s \neq t)$ are asymptotically independent;

8. $\sqrt{r_s}(\dot{\lambda}_s - \lambda_s)$ is asymptotically normal with expectation 0 and variance $2\lambda_s$.

The estimators $\dot{\lambda}$ can be *refined* in two simple ways. First, if the matrix V of (3.9) is non-singular – a sufficient condition is that the number of regressors k is smaller than $1/2 \min(r_1, r_2 \ldots r_m)$ – (cf. Streitberg, 1977, p. 83), a bias correction

$$\ddot{\boldsymbol{\lambda}} = V^{-1} \dot{\boldsymbol{\lambda}} \tag{3.12}$$

suggests itself. It can be shown (Streitberg, 1977, p. 86) that these estimators are local best in the class of all translation-invariant, unbiased, quadratic estimators of $\boldsymbol{\lambda}$. "Local best" means that $\ddot{\lambda}_s$ has minimum variance in this class of estimators, if the true parameter values coincide with the a priori-assumptions in (3.6). Alternatively, Rao's MINQUE-theory (cf. Rao, 1971) leads to $\ddot{\boldsymbol{\lambda}}$. Asymptotically V^{-1} is 'almost' an identity matrix:

$$V^{-1} = I + 0(\frac{1}{N}) \tag{3.13}$$

consequently $\dot{\boldsymbol{\lambda}}$ and $\ddot{\boldsymbol{\lambda}}$ are asymptotically equivalent estimators.
The second way to refine $\dot{\boldsymbol{\lambda}}$ is to use this estimate as starting point of an iteration procedure. Here, from $\dot{\boldsymbol{\lambda}}$ two-phase estimators of $\boldsymbol{\beta}$ are computed via (3.6), where the estimated covariance matrix is inserted instead of $\boldsymbol{\Sigma}_0$. This yields new residuals and updated estimates of $\boldsymbol{\lambda}$. Continuing this operation, estimators $\hat{\boldsymbol{\lambda}}, \hat{\boldsymbol{\beta}}$ are obtained which are solutions of the likelihood-equations. Indeed, the first partial derivatives of the log-likelihood-function

$$l(\underline{\boldsymbol{\beta}}, \underline{\boldsymbol{\lambda}}) = -\frac{1}{2} \log \det(\lambda_1 \boldsymbol{K}_1 + \ldots + \lambda_m \boldsymbol{K}_m) - \frac{1}{2} (\boldsymbol{y}-\boldsymbol{X\beta})'(\lambda_1 \boldsymbol{K}_1 + \ldots + \lambda_m \boldsymbol{K}_m)^{-1} (\boldsymbol{y}-\boldsymbol{X\beta}) \tag{3.14}$$

are simply

$$\frac{\partial l}{\partial \boldsymbol{\beta}} = \boldsymbol{X}'\boldsymbol{\Sigma}^{-1}\boldsymbol{y} - \boldsymbol{X}'\boldsymbol{\Sigma}^{-1}\boldsymbol{X\beta} \tag{3.15}$$

$$\frac{\partial l}{\partial \lambda_s} = -\frac{r_s}{2\lambda_s} + \frac{(\boldsymbol{y}-\boldsymbol{X\beta})'\boldsymbol{K}_s(\boldsymbol{y}-\boldsymbol{X\beta})}{2\lambda_s^2} \tag{3.16}$$

Equating these to zero, yields the equations

$$\hat{\boldsymbol{\beta}} = (\boldsymbol{X}'\hat{\boldsymbol{\Sigma}}^{-1}\boldsymbol{X})^{-1}\boldsymbol{X}'\hat{\boldsymbol{\Sigma}}^{-1}\boldsymbol{y} \tag{3.17}$$

$$\hat{\lambda}_s = r_s^{-1}(\boldsymbol{y}-\boldsymbol{X}\hat{\boldsymbol{\beta}})'\boldsymbol{K}_s(\boldsymbol{y}-\boldsymbol{X}\hat{\boldsymbol{\beta}}) \tag{3.18}$$

used alternately in the iteration procedure considered. Practical experience with the algorithm shows rapid convergence, quite regardless of starting values $\boldsymbol{\lambda}^{(0)}$. The algorithm breaks down if one of the iterates $\hat{\lambda}_s$ becomes zero. Again a sufficient condition to exclude this situation with probability 1 is given by $k < \frac{1}{2} \min(r_1, r_2 \ldots r_m)$.

It can be shown that $\hat{\boldsymbol{\beta}}$ and Gauss-Markov-estimators $(\boldsymbol{X}'\boldsymbol{\Sigma}^{-1}\boldsymbol{X})^{-1}\boldsymbol{X}'\boldsymbol{\Sigma}^{-1}\boldsymbol{y}$ are asymptotically equivalent estimators. To derive confidence limits for $\hat{\boldsymbol{\beta}}$ in large samples, therefore, the asymptotic covariance matrix $\mathrm{var}[\hat{\boldsymbol{\beta}}] = (\boldsymbol{X}'\boldsymbol{\Sigma}^{-1}\boldsymbol{X})^{-1}$ could be used. Also, $\hat{\boldsymbol{\lambda}}$ and $\dot{\boldsymbol{\lambda}}$ are asymptotically equivalent. The information matrix

$$-\mathrm{E}\left[\frac{\partial^2 l}{\partial(\boldsymbol{\beta},\boldsymbol{\lambda})\,\partial(\boldsymbol{\beta},\boldsymbol{\lambda})}\right] = \mathrm{diag}\left[\boldsymbol{X}'\boldsymbol{\Sigma}^{-1}\boldsymbol{X}, \frac{1}{2}\frac{r_1}{\lambda_1^2}, \ldots, \frac{1}{2}\frac{r_m}{\lambda_m^2}\right] \tag{3.19}$$

shows that $\hat{\boldsymbol{\beta}}$ and $\hat{\boldsymbol{\lambda}}$ are best asymptotically normal (BAN)-estimators of $\boldsymbol{\beta}$ and $\boldsymbol{\lambda}$.

6.3.2.4 *Estimators, using strategy 2*

Since $\boldsymbol{K}_s$ is idempotent of rank r_s, there exists an (N, r_s)-matrix $\boldsymbol{T}_s$ such that $\boldsymbol{K}_s = \boldsymbol{T}_s\boldsymbol{T}_s'$ and $\boldsymbol{T}_s'\boldsymbol{T}_s = \boldsymbol{I}$ (cf. section 6.2.3.2.). This suggests the following transformation to a *partial model* $\boldsymbol{y}_s = \boldsymbol{T}_s'\boldsymbol{y}$ with $\mathrm{E}[\boldsymbol{y}_s] = \boldsymbol{T}_s'\boldsymbol{X}\boldsymbol{\beta}$ and $\mathrm{var}[\boldsymbol{y}_s] = \lambda_s \boldsymbol{I}$. We obtain a standard linear model with r_s independent observations.

The parameter $\boldsymbol{\beta}$ may not be estimable in this partial model, since $\boldsymbol{T}_s'\boldsymbol{X}$ does not necessarily have full column rank and the normal equations

$$(\boldsymbol{X}'\boldsymbol{K}_s\boldsymbol{X})\boldsymbol{\beta} = \boldsymbol{X}'\boldsymbol{K}_s\boldsymbol{y} \tag{3.20}$$

are not uniquely solvable for $\boldsymbol{\beta}$. But residuals

$$\boldsymbol{e}_s = (\boldsymbol{I} - \boldsymbol{T}_s'\boldsymbol{X}(\boldsymbol{X}'\boldsymbol{K}_s\boldsymbol{X})^{-}\boldsymbol{X}'\boldsymbol{T}_s)\boldsymbol{T}_s'\boldsymbol{y} \tag{3.21}$$

are invariant under different choices of the generalized inverse $(\boldsymbol{X}'\boldsymbol{K}_s\boldsymbol{X})^{-}$ of $\boldsymbol{X}'\boldsymbol{K}_s\boldsymbol{X}$ (a matrix A^{-} is called a generalized inverse of A, when $AA^{-}A = A$). So quadratic forms

$$\boldsymbol{e}_s'\boldsymbol{e}_s = \boldsymbol{y}'\boldsymbol{B}\boldsymbol{y} \tag{3.22}$$

with $\boldsymbol{B} = \boldsymbol{K}_s - \boldsymbol{K}_s\boldsymbol{X}(\boldsymbol{X}'\boldsymbol{K}_s\boldsymbol{X})^{-}\boldsymbol{X}'\boldsymbol{K}_s$ can be used to derive an estimator of λ_s. Unbiased estimators exist if $tr[\boldsymbol{B}] > 0$, i.e. if $r_s > rg(\boldsymbol{X}'\boldsymbol{K}_s\boldsymbol{X})$. Subject to this condition we propose

$$\tilde{\lambda}_s = (tr\,\boldsymbol{B})^{-1}\,\boldsymbol{y}'\boldsymbol{B}\boldsymbol{y} \tag{3.23}$$

as an estimator of λ_s.

Essentially these estimators have been considered by Balestra (1973) in a model with two principal idempotents, but Balestra used $(\boldsymbol{X}'\boldsymbol{K}_s\boldsymbol{X})^{-1}$ in equation (3.21) (a matrix that does not necessarily exist) and tried to derive $\tilde{\lambda}_s$ as a uniformly best unbiased quadratic estimator, which is, in general, wrong. Properties of $\tilde{\lambda}_s$ follow immediately from standard linear model theory:

1. $\tilde{\lambda}_s$ exists and is non-negative if $tr\,\boldsymbol{B} > 0$;
2. $\tilde{\lambda}_s$ is translation invariant;
3. A multiple of $\tilde{\lambda}_s$ is chi square-distributed, i.e.

$$\frac{b_s}{\lambda_s}\,\tilde{\lambda}_s \sim \chi^2\,[b_s] \text{ where } b_s = tr\,\boldsymbol{B};$$

4. $E[\tilde{\lambda}_s] = \lambda_s$;
5. $\operatorname{var}[\tilde{\lambda}_s] = \dfrac{2\lambda_s^2}{b_s}$;
6. $\tilde{\lambda}_s$ and $\tilde{\lambda}_t$ are independent if $s \neq t$;
7. $\tilde{\lambda}_s$ and $\dot{\boldsymbol{\beta}}$ (of (3.6)) are independent.

Normality has been assumed in (3), (5), (6), (7). Asymptotically we find

8. $\tilde{\lambda}_s$ is consistent;
9. $\sqrt{r_s}\,(\tilde{\lambda}_s - \lambda_s)$ is asymptotically normal with expectation 0 and variance $2\lambda_s$.

From (9) follows that all estimators of λ_s considered so far are BAN-estimators.

It may be instructive to look at two examples, where we assume $\boldsymbol{X} = \boldsymbol{\iota}$.

Example 1. Stationary processes:

The covariance matrix is $\boldsymbol{\Sigma} = \lambda_0\boldsymbol{K}_0 + \lambda_1\boldsymbol{K}_1 + \ldots \lambda_M\boldsymbol{K}_M$, where $\boldsymbol{K}_s$ is the projector corresponding to frequency $\frac{s}{N}$. From the definition of $\boldsymbol{K}_s$ we find:

(1) $\boldsymbol{K}_0 \boldsymbol{\iota} = \boldsymbol{\iota}$; (2) $tr\, \boldsymbol{K}_0 = 1$; (3) $\boldsymbol{K}_s \boldsymbol{\iota} = \boldsymbol{0}$; (4) $tr\, \boldsymbol{K}_s = 2$, for $s \neq 0$.

λ_0 cannot be estimated by the method proposed. For $s \neq 0$ we find

$$\tilde{\lambda}_s = \frac{1}{2} \boldsymbol{y}' \boldsymbol{K}_s \boldsymbol{y}.$$

Since $\boldsymbol{K}_s = (K^{(s)}_{\alpha\beta}) = (\frac{2}{N} \cos 2\pi \frac{s}{N} (\alpha-\beta))$ this is equal to

$$\tilde{\lambda}_s = \frac{1}{N} \sum_{\alpha=-M}^{M} \sum_{\beta=-M}^{M} y_\alpha y_\beta \cos 2\pi \frac{s}{N} (\alpha-\beta)$$

$$= \frac{1}{N} (\sum_\alpha y_\alpha \cos 2\pi \frac{s}{N} \alpha)^2 + \frac{1}{N} (\sum_\alpha y_\alpha \sin 2\pi \frac{s}{N} \alpha)^2 ,$$

the well-known periodogram estimator of λ_s. Note that the asymptotic assumptions are not justified for that particular dependency structure.

Example 2. Transaction tables:

Consider the 'reciprocal dependencies' model of section 6.2.3.6:

$$\boldsymbol{f} = \text{vec}\, \boldsymbol{F} \sim N[\boldsymbol{\mu} \boldsymbol{\iota}, \lambda_1 \frac{1}{2} (\boldsymbol{I}+\boldsymbol{\Phi}) + \lambda_2 \frac{1}{2} (\boldsymbol{I}-\boldsymbol{\Phi})],$$

where one finds: (1) $\boldsymbol{K}_1 \boldsymbol{\iota} = \boldsymbol{\iota}$; (2) $tr\, \boldsymbol{K}_1 = \frac{n(n+1)}{2}$; (3) $\boldsymbol{K}_2 \boldsymbol{\iota} = \boldsymbol{0}$; (4) $tr\, \boldsymbol{K}_2 = \frac{n(n-1)}{2}$. The partial model estimators are

$$\tilde{\lambda}_1 = \frac{\boldsymbol{f}' \boldsymbol{K}_1 (\boldsymbol{I} - \boldsymbol{\iota} (\boldsymbol{\iota}' \boldsymbol{\iota})^{-1} \boldsymbol{\iota}') \boldsymbol{K}_1 \boldsymbol{f}}{\frac{1}{2} n(n+1) - 1}$$

$$\tilde{\lambda}_2 = \frac{\boldsymbol{f}' \boldsymbol{K}_2 \boldsymbol{f}}{\frac{1}{N} n(n-1)} .$$

These estimators can be cast in an easily interpretable form, if we use the decomposition of $\boldsymbol{F}$ in a symmetric table $\boldsymbol{S} = \frac{1}{2}(\boldsymbol{F}+\boldsymbol{F}')$ and an antisymmetric part $\boldsymbol{A} = \frac{1}{2}(\boldsymbol{F}-\boldsymbol{F}')$. Denoting the (empirical) variances over all elements of $\boldsymbol{S}$ and $\boldsymbol{A}$ as s_S^2 and s_A^2 respectively, we see that $\tilde{\lambda}_1$ and $\tilde{\lambda}_2$ are multiples of these variances:

$$\tilde{\lambda}_1 = s_S^2 \left[\frac{2n^2}{n(n+1)-2}\right]$$

$$\tilde{\lambda}_2 = s_A^2 \left[\frac{2n}{n-1}\right].$$

6.3.2.5 *Estimation in type (B) models*

The complete sufficiency of $(\boldsymbol{y}'\boldsymbol{K}_s\boldsymbol{y}, \boldsymbol{X}'\boldsymbol{K}_s\boldsymbol{y})_{s=1,2\ldots m}$ for $(\lambda_s, \boldsymbol{\beta}_s)$ renders estimation of these parameters trivial. Any function of the complete sufficient statistic is the unique minimum variance unbiased estimator of its expectation. Since the density function factorizes into m independent partial models for $\boldsymbol{T}_1'\boldsymbol{y}, \boldsymbol{T}_2'\boldsymbol{y}, \ldots \boldsymbol{T}_m'\boldsymbol{y}$ with different parameter sets, the *standard linear model theory* can be used to see that $\tilde{\lambda}_s$ estimates λ_s and

$$\tilde{\boldsymbol{\beta}}_s = (\boldsymbol{X}'\boldsymbol{K}_s\boldsymbol{X})^{-1}\boldsymbol{X}'\boldsymbol{K}_s\boldsymbol{y} \qquad (3.24)$$

estimates $\boldsymbol{\beta}_s$ if the inverse of $\boldsymbol{X}'\boldsymbol{K}_s\boldsymbol{X}$ exists. If $\boldsymbol{X}'\boldsymbol{K}_s\boldsymbol{X}$ is singular, unbiased estimation of $\boldsymbol{\beta}_s$ is impossible, but in general certain linear combinations of the elements in $\boldsymbol{\beta}_s$ can be estimated.

6.3.3 *Remarks on test procedures and generalizations*

The paper focusses on estimation procedures, but for practical application of the model tests are necessary. Therefore, some short remarks on possible test procedures follow.

A test on independence. The model $(\boldsymbol{y}, \boldsymbol{X\beta}, \Sigma\, \lambda_s\boldsymbol{K}_s)$ becomes a standard linear model with independent observations $(\boldsymbol{y}, \boldsymbol{X\beta}, \lambda\boldsymbol{I})$ if all λ_s are equal to λ. A likelihood-ratio-statistic for the problem

$H_0 : \lambda_1 = \lambda_2 = \ldots = \lambda_m$

H_1 : not all λ_s equal

is easily derived. Let $L(\underline{\boldsymbol{\beta}}, \lambda_1, \ldots, \lambda_m)$ be the likelihood-function of the model with dependent observations and $\hat{\boldsymbol{\beta}}, \hat{\lambda}_1 \ldots \hat{\lambda}_m$ the corresponding ML-estimators. Under H_0 ML-estimators of the parameters $\boldsymbol{\beta}$, λ are $\hat{\boldsymbol{\beta}}_0 = (X'X)^{-1}X'y$ and $\hat{\lambda} = N^{-1}(y-X\hat{\boldsymbol{\beta}}_0)'(y-X\hat{\boldsymbol{\beta}}_0)$. The LR-statistic is therefore

$$\mathcal{L} = \frac{\sup_{\boldsymbol{\beta},\lambda} L(\boldsymbol{\beta}, \lambda, \lambda \ldots \lambda)}{\sup_{\boldsymbol{\beta}, \lambda_1 \ldots \lambda_m} L(\boldsymbol{\beta}, \lambda_1, \lambda_2 \ldots \lambda_m)} = \frac{L(\hat{\boldsymbol{\beta}}_0, \hat{\lambda}, \ldots, \hat{\lambda})}{L(\hat{\boldsymbol{\beta}}, \hat{\lambda}_1, \ldots, \hat{\lambda}_m)} = \prod_{s=1}^{m} \left(\frac{\hat{\lambda}}{\hat{\lambda}_s}\right)^{\frac{r_s}{2}} \tag{3.25}$$

If the asymptotic assumptions outlined in 6.3.2.2 are justified

$$T_1 = -2 \log \mathcal{L} = -N \log \hat{\lambda} + \sum_{s=1}^{m} r_s \log \hat{\lambda}_s \tag{3.26}$$

is asymptotically distributed as a $\chi^2 [m-1]$-variate.

This test can be further refined. The test problem is (after an orthogonal transformation) formally equivalent to the well-known test on equality of m variances, so we could use the independent and chi square-distributed partial model estimators

$$\tilde{\lambda}_s \sim \frac{\lambda_s}{b_s} \chi^2 [b_s] \tag{3.27}$$

and the following "pooled" estimator

$$\lambda^* = \left(\sum_{s=1}^{m} b_s\right)^{-1} \left(\sum_{s=1}^{m} b_s \tilde{\lambda}_s\right), \tag{3.28}$$

in Bartlett's modified LR-statistic:

$$T_2 = [1 + (3m-3)^{-1} \left(\sum_{s=1}^{m} \frac{1}{b_s} \right.$$

$$-\left(\sum_{s=1}^{m} \frac{1}{b_s}\right)^{-1}\Big)\Big]^{-1} \Big[-\Big(\sum_{s=1}^{m} b_s\Big)^{-1} \tag{3.29}$$

$$\log \lambda^* + \sum_{s=1}^{m} b_s \log \tilde{\lambda}_s \Big],$$

being (under H_0) approximately $\chi^2\,[m - 1]$-distributed (even for small b_s !). This test is known to be rather nonrobust against deviations from normality and robust tests on homogenity of variances could be adapted to the present case (cf. Layard, 1973; Levene, 1960).

Test on substructures. If $(\boldsymbol{K}_1, \boldsymbol{K}_2 \ldots \boldsymbol{K}_m)$ defines a decomposable covariance structure, substructures can be obtained by 'melting' together certain of the $\boldsymbol{K}_i$, e.g. $(\boldsymbol{K}_1 + \boldsymbol{K}_2 + \ldots + \boldsymbol{K}_a, \boldsymbol{K}_{a+1}, \ldots \boldsymbol{K}_m)$ is also a decomposable structure, where certain 'dependency strata' are identified. To test for instance $H_0 : \lambda_1 = \lambda_2 = \ldots = \lambda_a$ against general alternatives again Bartlett's statistic can be utilized, when m in (3.28) and (3.29) is replaced by a.

A test on *equality of eigenvalues in groups* can be obtained in the same way. Let $\pi = (Z_1, Z_2 \ldots Z_G)$ constitute a partition of $\{1, 2 \ldots m\}$ in G sets Z_g. We want to test

$$H_0 : \lambda_s = \lambda_t \text{ if } s \text{ and } t \text{ belong to the same } Z_g$$

against a general alternative. Defining the pooled estimator in group Z_g by

$$\lambda_g^* = \frac{1}{B_g} \sum_{s \in Z_g} b_s \tilde{\lambda}_s \text{ where } B_g := \sum_{s \in Z_g} b_s \tag{3.30}$$

we can, by expanding the expectation of a modified LR-statistic

$$T_3 = -\sum_{g=1}^{G} B_g \log \lambda_g^* + \sum_{s=1}^{m} b_s \log \tilde{\lambda}_s \tag{3.31}$$

up to order d_s^{-2}, obtain the test statistic

$$T_4 = [1 + (3m-3G)^{-1} (\sum_{s=1}^{m} \frac{1}{b_s} - \sum_{g=1}^{G} \frac{1}{B_g})]^{-1} T_3, \tag{3.32}$$

distributed asymptotically as χ^2 with $m - G$ degrees of freedom under H_0.

Tests on $\boldsymbol{\beta}$. For large samples the asymptotic equivalence of the ML-estimator for $\boldsymbol{\beta}$ and the Gauss-Markov-estimator could be utilized to derive tests or confidence bounds about $\boldsymbol{\beta}$. For small samples, one should recognize the fact that tests on $\boldsymbol{\beta}$ are essentially Behrens-Fisher problems and adopt the known partial solutions (e.g. James, 1954).

Generalizations:

1. Certain assumptions about the dependency structure (cf. 6.2.3.7 and 6.2.3.8) lead to models with linear constraints among the eigenvalue parameters λ_s. Along the lines of Aitchison and Silvey (1958, 1960) estimation procedures for these models have been developed and will be presented elsewhere.
2. One of the editors noted that the proposed procedure could be used to estimate spatial AR and ARMA processes. We hope to explore this possibility and compare it with alternative approaches in a paper to follow.

6.3.4 *Multivariate linear models*

Estimation problems in multivariate linear models $(\boldsymbol{Y}, \boldsymbol{XB}, \Sigma \boldsymbol{K}_s \otimes \boldsymbol{\Omega}_s)$, where $\mathrm{E}[\boldsymbol{Y}] = \boldsymbol{XB}$ and $\mathrm{var}[\mathrm{vec}\ \boldsymbol{Y}] = \Sigma \boldsymbol{K}_s \otimes \boldsymbol{\Omega}_s$, are rather similar to corresponding univariate problems. We confine ourselves to some short remarks. The loglikelihood function of the model is

$$l(\boldsymbol{B}, \boldsymbol{\Omega}_1, \ldots, \boldsymbol{\Omega}_m) = -\frac{1}{2} \log \det (\sum_{s=1}^{m} \boldsymbol{K}_s \otimes \boldsymbol{\Omega}_s)$$

$$- \frac{1}{2} (\mathrm{vec}\, \boldsymbol{Y} - (\boldsymbol{X} \otimes \boldsymbol{I})$$

$$\mathrm{vec}\, \boldsymbol{B})' (\sum_{s=1}^{m} \boldsymbol{K}_s \otimes \boldsymbol{\Omega}_s)^{-1}$$

$$(\text{vec}\, \boldsymbol{Y} - (\boldsymbol{X} \otimes \boldsymbol{I})\, \text{vec}\, \boldsymbol{B})$$

$$= -\frac{1}{2} \sum_{s=1}^{m} r_s \log \det \boldsymbol{\Omega}_s$$

$$-\frac{1}{2} \sum_{s=1}^{m} tr\,(\boldsymbol{Y}-\boldsymbol{XB})'\boldsymbol{K}_s(\boldsymbol{Y}-\boldsymbol{XB})\boldsymbol{\Omega}_s^{-1}$$

(3.33)

with partial derivatives

$$\frac{\partial l}{\partial \boldsymbol{B}} = \sum_{s=1}^{m} (\boldsymbol{X}'\boldsymbol{K}_s \boldsymbol{Y} \boldsymbol{\Omega}_s^{-1} - \boldsymbol{X}'\boldsymbol{K}_s \boldsymbol{X}\boldsymbol{B}\boldsymbol{\Omega}_s^{-1}) \quad (3.34)$$

$$\frac{\partial l}{\partial \boldsymbol{\Omega}_s} = \frac{1}{2} \boldsymbol{\Omega}_s^{-1} (\boldsymbol{Y}-\boldsymbol{XB})'\boldsymbol{K}_s (\boldsymbol{Y}-\boldsymbol{XB})\boldsymbol{\Omega}_s^{-1} - \frac{1}{2} r_s \boldsymbol{\Omega}_s^{-1}. \quad (3.35)$$

ML-estimators can be found in the same way as in (3.17), (3.18) by an iteration procedure, cycling between the resulting likelihood-equations

$$\text{vec}\, \hat{\boldsymbol{B}} = [\sum_{s=1}^{m} \boldsymbol{X}'\boldsymbol{K}_s\boldsymbol{X} \otimes \boldsymbol{\Omega}_s^{-1}]^{-1} [\sum_{s=1}^{m} \boldsymbol{X}'\boldsymbol{K}_s \otimes \boldsymbol{\Omega}_s^{-1}]\, \text{vec}\, \boldsymbol{Y}$$

(3.36)

$$\hat{\boldsymbol{\Omega}}_s = r_s^{-1} (\boldsymbol{Y}-\boldsymbol{X}\hat{\boldsymbol{B}})'\boldsymbol{K}_s(\boldsymbol{Y}-\boldsymbol{X}\hat{\boldsymbol{B}}). \quad (3.37)$$

In general, if no particular a priori information were available, one would use $\hat{\boldsymbol{\Omega}}_s = \mathbf{I}$ as a starting point of the iteration procedure, thus obtaining the standard OLS-estimators as first iterates. The first step of the algorithm yields estimators $\dot{\boldsymbol{B}}, \dot{\boldsymbol{\Omega}}_s$ corresponding to $\dot{\boldsymbol{\beta}}, \dot{\boldsymbol{\lambda}}_s$ of the univariate model.

A third class of estimators which can easily be generalized to the multivariate case are 'partial model' estimators $\tilde{\lambda}_s$. We find

$$\tilde{\boldsymbol{\Omega}}_s = \frac{Y'(K_s - K_s X(X'K_s X)^- X'K_s)Y}{b_s} \tag{3.38}$$

where $b_s = tr\, K_s - rg\, X'K_s X$.

Note that $\tilde{\boldsymbol{\Omega}}_s$ is central Wishart-distributed with degrees of freedom b_s and covariance-parameter $b_s \boldsymbol{\Omega}_s$.

The extension of Minque-estimators to the multivariate case is more difficult and will not be considered in this paper, because the computational problems are quite horrible. In general, a linear equation system of order $N^2 + m$ has to be solved and only for very special types of a-priori-assumptions (e.g. all covariance-matrices $\boldsymbol{\Omega}_s$ equal up to scale-parameters) the univariate solution of Rao is obtained.

6.4 PRINCIPAL COMPONENTS

In principal components analysis, we seek a linear combination $\boldsymbol{a}_1' \boldsymbol{y}$ of a random vector $\boldsymbol{y} \sim N[\boldsymbol{\mu}, \boldsymbol{\Sigma}]$ having maximum variance $\text{var}[\boldsymbol{a}_1'\boldsymbol{y}] = \boldsymbol{a}_1' \boldsymbol{\Sigma} \boldsymbol{a}_1$ among all linear combinations $\boldsymbol{a}_1'\boldsymbol{a}_1 = 1$. Further principal components $\boldsymbol{a}_2'\boldsymbol{y} \ldots \boldsymbol{a}_p'\boldsymbol{y}$ are defined in the same way, subject to the condition of being orthogonal to all preceding ones. Optimal weight vectors $\boldsymbol{a}$, together with variances λ of their principal components, are found by solving the resulting eigenproblem $\boldsymbol{\Sigma a} = \lambda \boldsymbol{a}$. Sample principal components are generated analogously from the eigenproblem $\boldsymbol{Sa} = \lambda \boldsymbol{a}$, where $\boldsymbol{S}$ is an estimate of $\boldsymbol{\Sigma}$ obtained from N independent observations on $\boldsymbol{y}$.

Assuming dependencies among the rows of $\boldsymbol{Y}$, we are given not N observations on the same random vector but one observation on a random matrix. One could consider several approaches to define principal components in this case, but due to space limitations only one will be presented here.

This approach is strongly tied to the particular dependency model. We define the first principal component as a 'filtered' version of $\boldsymbol{Y}$:

$$\boldsymbol{f}_1 = w_1 \boldsymbol{K}_1 \boldsymbol{Y} \boldsymbol{a}_1 + w_2 \boldsymbol{K}_2 \boldsymbol{Y} \boldsymbol{a}_2 + \ldots + w_m \boldsymbol{K}_m \boldsymbol{Y} \boldsymbol{a}_m \tag{4.1}$$

i.e. a weighted sum of contributions from the different levels of the dependency structure, where the weights w_s can be chosen by the user to emphasize or deemphasize certain levels in the structure. If no levels are to be particularly attenuated, one would set $w_s = 1$ for all s. The variance-covariance matrix of $\boldsymbol{f}_1$ is

$$\operatorname{var}[\boldsymbol{f}_1] = \sum_{s=1}^{m} w_s^2 (\boldsymbol{a}_s' \boldsymbol{\Omega}_s \boldsymbol{a}_s) \boldsymbol{K}_s \tag{4.2}$$

so $\boldsymbol{f}_1$ follows the same type of dependency structure as the columns of $\boldsymbol{Y}$.

Again we could use the trace of this matrix as criterion to maximize, subject to $\boldsymbol{a}_s'\boldsymbol{a}_s = 1$. This leads to the solution of m eigenvalue problems

$$\boldsymbol{\Omega}_s \boldsymbol{a}_s = \lambda \boldsymbol{a}_s \tag{4.3}$$

from where the weight vectors that generate further principal components can also be found.

It is remarkable that this result is not changed if we use other sensible scalar measures of the 'variance' of $\boldsymbol{f}_1$ based on $\operatorname{var}[\boldsymbol{f}_1]$ like quadratic forms $\boldsymbol{c}'\operatorname{var}[\boldsymbol{f}_1]\boldsymbol{c}$, $\det \operatorname{var}[\boldsymbol{f}_1]$, the j'th largest eigenvalue of $\operatorname{var}[\boldsymbol{f}_1]$, etc. To show this, let $\boldsymbol{f}_1^*$ be any other vector having the form (4.1) with weight vectors $\boldsymbol{a}_s^*$, fulfilling $\boldsymbol{a}_s^{*\prime}\boldsymbol{a}_s^* = 1$. The difference between $\operatorname{var}[\boldsymbol{f}_1]$ and $\operatorname{var}[\boldsymbol{f}_1^*]$ is

$$\operatorname{var}[\boldsymbol{f}_1] - \operatorname{var}[\boldsymbol{f}_1^*] = \sum_{s=1}^{m} w_s^2 (\boldsymbol{a}_s' \boldsymbol{\Omega}_s \boldsymbol{a}_s - \boldsymbol{a}_s^{*\prime} \boldsymbol{\Omega}_s \boldsymbol{a}_s^*) \boldsymbol{K}_s \tag{4.4}$$

For all possible choises of $\boldsymbol{a}_s$ this matrix is non-negative definite, since the first eigenvector $\boldsymbol{a}_s$ of $\boldsymbol{\Omega}_s$ maximizes the quadratic form $\boldsymbol{a}_s' \boldsymbol{\Omega}_s \boldsymbol{a}_s$. The above statement follows immediately from this observation. If $\boldsymbol{\Omega}_s$ is estimated by $\tilde{\boldsymbol{\Omega}}_s$, eigenvectors $\boldsymbol{a}_s^*$ of different $\tilde{\boldsymbol{\Omega}}_s$ are independent, and the asymptotic distribution of eigenvectors and eigenvalues of $\tilde{\boldsymbol{\Omega}}_s$ is well-known (cf. Girshick, 1939; Anderson, 1963).

6.5 CONCLUSION

Three large areas of multivariate analysis have not yet been covered: canonical analysis, discriminant analysis, and cluster analysis. While canonical analysis can easily be generalized to systems of dependent observations in largely the same way as principal components (one partitions $\boldsymbol{Y}$ in $[\boldsymbol{X}|\boldsymbol{Z}]$ and seeks 'filtered' versions $\Sigma\ \boldsymbol{K}_s\boldsymbol{X}\boldsymbol{a}_s$ and $\Sigma\ \boldsymbol{K}_s\boldsymbol{Z}\boldsymbol{b}_s$ such that a criterion applied to the covariance-matrix of these vectors is maximized), the adaptation of discriminant analysis and cluster methods is wrought with interpretational problems. The following remarks are rather speculative.

The methods considered so far all deal with the 'columns' aspect of the data matrix $\boldsymbol{Y}$: the main interest of principal components, linear models, canonical analysis lies in the analysis of the interrelations between variables.

Discriminant analysis and cluster analysis, on the other hand, are 'observation oriented': these methods view the data matrix $\boldsymbol{Y}$ as a collection of "rows" and express their results in terms of individual observations (e.g. "row $\boldsymbol{x}$ belongs to group 1", "rows $\boldsymbol{a}$, $\boldsymbol{b}$, $\boldsymbol{c}$ constitute one cluster", etc.). Now, if the observations are dependent, they loose a certain amount of their individuality (this is problematic for cluster analysis) and nearly all of their exchangeability (so it matters in which row of $\boldsymbol{Y}$ the vector "$\boldsymbol{x}$" stands if the problem of assigning $\boldsymbol{x}$ to a group of observations is given).

I have not yet found a sensible way of defining cluster analysis for dependent observations but it is stimulating to think about alternative ways to do it (should we cluster the independent transformed observations in $\boldsymbol{T}'\boldsymbol{Y}$; should we restrict clusters to certain 'obvious' sets of observations like time segments; should we adapt a stochastic mixture algorithm?).

A participant of the symposium suggested that m different cluster analyses should be done for the m different levels $\boldsymbol{K}_1\boldsymbol{Y}$, $\boldsymbol{K}_2\boldsymbol{Y} \ldots \boldsymbol{K}_m\boldsymbol{Y}$ of the dependency structure.

I think (at least certain aspects) of discriminant analysis could be generalized fairly easily. It is instructive to look at a particular (simplified) example:

The city of West Berlin shows a rather clear-cut east-west division: working-class districts in the east and middle-class districts in the west. While most districts can be unequivocally put into the "east" or the "west" group, let us assume there is one district we cannot reach a simple decision about. To which of the two groups does it belong, if a certain number of socio-economic characteristics are considered?

If we look at this example more closely, we see that it is essentially a multivariate linear problem with a "somewhat" unknown design matrix. Assume, for purposes of exposition, that Berlin had only

5 districts, where district 3's designation is unknown and (1, 2) belong to the east group, (4, 5) to the west group. The problem then is to decide between two alternative models $(\boldsymbol{Y}, \boldsymbol{X}_1\boldsymbol{B}, \Sigma \boldsymbol{K}_s \otimes \boldsymbol{\Omega}_s)$ and $(\boldsymbol{Y}, \boldsymbol{X}_2\boldsymbol{B}, \Sigma \boldsymbol{K}_s \otimes \boldsymbol{\Omega}_s)$ where

$$\boldsymbol{X}_1 = \begin{bmatrix} 1 & 0 \\ 1 & 0 \\ 1 & 0 \\ 0 & 1 \\ 0 & 1 \end{bmatrix} \qquad \text{and } \boldsymbol{X}_2 = \begin{bmatrix} 1 & 0 \\ 1 & 0 \\ 0 & 1 \\ 0 & 1 \\ 0 & 1 \end{bmatrix}$$

Thus formalized, the problem has an intuitively obvious solution. We analyze the data with both of the models at stake and take the (maximized) likelihoods L_1 and L_2 under model 1 (resp. 2) as decision criterion: if $L_1 > L_2$, accept model 1, otherwise model 2 (a priori probabilities assumed equal).

This approach can of course be extended to more than two groups and more than one observation left undecided – we simply have to consider a larger set of possible design matrices. Note that cluster analysis could be viewed as an extreme case of this, when the set of possible design matrices consists of all (N, k)-matrices specifying a partition of the original observations in k groups. It is intriguing to speculate about this idea of discriminant analysis or cluster analysis as multivariate linear models with (partially) unknown design matrices and try a generalization to more interesting classes of $\boldsymbol{X}$-matrices than just single-classification designs.

Back to the dependency structures handled in the paper. The proof of the pudding lies in the eating – the proof of statistical methods lies in their application. I could close as Besag did in a fine paper of him (Besag, 1976) and plead for data analysis, but I fear there is a lot to be done *before* actual data analysis with the model proposed could be recommended. So let me end the paper by pointing out some – not all – of the points requiring further thought:

1. How robust are the procedures against distributional deviations in the direction of "long tail" distributions? Or, since nonrobustness is quite obvious if one uses dubious statistics like

means, variances, covariances: how could one robustify estimators and test procedures?
2. What is the effect of a wrong (and we are sure to be wrong here) specification of the particular dependency structure?
3. How should one compare the merits of two alternative dependency structures in explaining a given set of data?
4. How does one approximate a general 'linear' dependency structure of the form $\text{var}[y] = \alpha_1 A_1 + \alpha_2 A_2 + \ldots + \alpha_m A_m$ by a decomposable one (cf. the approximation of stationary processes by circular ones)? Or should we have used the larger class from the beginning?

REFERENCES

Aitchison, J. and Silvey, S. D. (1958), 'Maximum likelihood estimation of parameters subject to restraints', *Annals of Mathematical Statistics*, 29, 813-828.

Aitchison, J. and Silvey, S. D. (1960), 'Maximum likelihood estimation procedures and associated tests of significance', *Journal of the Royal Statistical Society*, B, 22, 154-171.

Anderson, T. W. (1963), 'Asymptotic theory for principal component analysis', *Annals of Mathematical Statistics*, 34, 122-148.

Anderson, T. W. (1971), *Statistical analysis of time-series*, Wiley, New York.

Bartels, C. P. A. (1976), 'On the power of the generalized Moran contiguity coefficient in testing for spatial autocorrelation among regression disturbances', COMPSTAT, 181-188.

Balestra, P. (1973), 'Best quadratic unbiased estimators of the variance-covariance matrix in normal regression', *Journal of Econometrics*, 1, 17-28.

Bellman, R. (1960), *Introduction to matrix analysis*, McGraw-Hill, New York.

Berry, B. L. J. (1971), 'Problems of data organisation and analytical methods in geography', *Journal of the American Statistical Association*, 66, 510-523.

Besag, J. (1975), 'Statistical analysis of non-lattice data', *The Statistician*, 24, 179-197.

Clark, D., Davies, W. D. D. and Johnston, R. J. (1974), 'The application of factor analysis in human geography', *The Statistician*, 23, 259-281.

Cliff, A. D. and Ord, J. K. (1972), 'Testing for spatial autocorrelations among regression residuals', *Geographical Analysis*, 4, 267-284.

Cliff, A. D. and Ord, J. K. (1975), 'Model building and the analysis of spatial pattern in human geography', *Journal of the Royal Statistical Society, B*, 37, 297-348.

Cliff, A. D., Haggett, P., Ord, J. K., Bassett, K. and Davies, B. E. (1975), *Elements of spatial structure*, Cambridge University Press, Cambridge.

Durbin, J. and Watson, G. S. (1971), 'Testing for serial correlation in least squares regression III', *Biometrika*, 58, 1-19.

Eaton, M. (1972), *Introduction to multivariate statistical methods*, Copenhagen.

Fisher, W. D. (1958), 'On grouping for maximal homogenity', *Journal of the American Statistical Association*, 53, 789-798.

Fuller, W. A. (1976), *Introduction to statistical time-series*, Wiley, New York.

Girschick, M. A. (1939), 'On the sampling theory of roots of determinantal equations', *Annals of Mathematical Statistics*, 10, 203-224.

Granger, C. W. J. (1975), 'Aspects of the analysis and interpretation of temporal and spatial data', *The Statistician*, 24, 197-210.

Graybill, F. A. and Huntquist, R. A. (1961), 'Theorems concerning Eisenhardt's Model 2', *Annals of Mathematical Statistics*, 32, 261-296.

Hartley, H. O. and Rao, J. N. K. (1967), 'Maximum likelihood estimation for the mixed analysis of variance model', *Biometrika*, 54, 93-108.

James, G. S. (1954), 'Tests of linear hypotheses in univariate and multivariate analysis when the ratios of population variances are unknown', *Biometrika*, 38, 19-43.

Kuhbier, P. (1976), *Linear decomposition models I*, Discussion paper, Institut für Quantitative Ökonomie und Statistik, FU Berlin.

Lehmann, E. L. (1959), *Testing statistical hypotheses*, Wiley, New York.

Mather, P. M. and Openshaw, S. (1974), 'Multivariate methods and geographical data', *The Statistician*, 23, 283-308.

Nissen, D. H. (1968), 'A note on the variance of a matrix', *Econometrica*, 36, 603-604.

Rao, C. R. (1971), 'Minimum variance quadratic unbiased estimation of variance components', *Journal of Multivariate Analysis*, 1, 445-456.

Streitberg, B. (1977), *Schätzung von Kovarianzstrukturen in linearen Zwei- und Mehrebenenmodellen*, FU Berlin.

Unwin, D. J. and Hepple, L. W. (1974), 'The statistical analysis of spatial series', *The Statistician*, 23, 211-217.

Acknowledgement: For helpful discussions, thanks go to N. Apostolopoulos, M. Bless, R. Edmondson. I am especially thankful to M. Bless who fulfilled the heroic task of typing all the horrible formulae.

7. Bayesian analysis of the linear model with spatial dependence

LESLIE W. HEPPLE*

7.1 INTRODUCTION

This paper develops a Bayesian analysis of the linear regression model with spatial dependence in the disturbances. The existing literature in spatial econometrics has been based entirely on sampling theory approaches (e.g. Neyman-Pearson and maximum likelihood) to statistical inference, and the Bayesian perspective has not been explored for spatial estimation. In contrast, the mainstream of econometric work has been influenced by the Bayesian approach to statistical inference during the last ten years, largely through the texts and papers of Zellner and Box (Zellner, 1971; Box and Tiao, 1973) and their associates and students. Bayesian methods have been fruitfully applied to the analysis and estimation of a wide range of econometric problems, such as simultaneous equations (Chetty, 1968), production functions (Tsurumi and Tsurumi, 1976; Zellner and Richard, 1973), distributed lag models (Zellner and Geisel, 1970), the linear model with serially correlated errors (Zellner and Tiao, 1965), and the linear model with non-normal errors (Box and Tiao, 1962).

Two attractive features of the Bayesian approach are the ability to make exact, finite-sample inferences in contexts and models where the sampling theory approach can only give large-sample, asymptotic inferences (see section 7.2), and the direct way it facilitates the analysis of robustness of models to specification error. The present paper sets spatial econometric work within this framework, using one specific model as example. The structure of the paper is as follows. Section 7.2 outlines some of the basic concepts of Bayesian inference; section 7.3 develops the Bayesian analysis

* University of Bristol, Department of Geography, Bristol, England.

of the linear model with spatial dependence in the disturbances; section 7.4 presents an empirical application of these techniques, and section 7.5 gives concluding remarks.

7.2 THE NATURE OF BAYESIAN INFERENCE

This section outlines some of the main ideas in Bayesian inference. This can only be a very brief account, and full explanations can be found in Zellner (1971) and Box and Tiao (1973).

In classical statistical inference (both Neyman-Pearson and maximum likelihood approaches), the observed data are assumed to be a random sample from a population of possible samples, and the sample is used to make inferences about population parameters, which are assumed fixed. In contrast, the Bayesian approach takes the observed data (assumed to come from a probability distribution of known general structure) and uses them to modify prior ideas about the parameters, which are assumed to have a probability distribution.

The central element in the Bayesian approach is the use of the information from the observed data to modify a prior probability distribution for the parameters into a posterior probability distribution through the medium of Bayes' Theorem or the Principle of Inverse Probability. To illustrate this, consider a vector of N observations, denoted by $\boldsymbol{y}$, whose probability distribution $p(\boldsymbol{y}|\boldsymbol{\beta})$ depends on the values of k parameters $\boldsymbol{\beta}$. Let the parameters $\boldsymbol{\beta}$ have a probability distribution $p(\boldsymbol{\beta})$ which expresses our prior knowledge of the probability distribution of $\boldsymbol{\beta}$ without any knowledge of the data. Then

$$p(\boldsymbol{y}\,|\boldsymbol{\beta})\,p(\boldsymbol{\beta}) = p(\boldsymbol{\beta}\,|\boldsymbol{y})\,p(\boldsymbol{y})$$

Once we are given the observed data $\boldsymbol{y}$, the conditional distribution of $\boldsymbol{\beta}$ is

$$p(\boldsymbol{\beta}\,|\boldsymbol{y}) = \frac{p(\boldsymbol{y}\,|\boldsymbol{\beta})\,p(\boldsymbol{\beta})}{p(\boldsymbol{y})}, \text{ if } p(\boldsymbol{y}) \neq 0 \tag{2.1}$$

That is, the combination of $p(\boldsymbol{\beta})$, the prior probability distribution of $\boldsymbol{\beta}$, with the information from the observed data, $p(\boldsymbol{y}\,|\boldsymbol{\beta})$, gives the posterior distribution of $\boldsymbol{\beta}$ given $\boldsymbol{y}$. The denominator of (2.1) can be obtained in practice as follows:

$$p(\mathbf{y}) = c^{-1} = \int p(\mathbf{y} \mid \boldsymbol{\beta}) p(\boldsymbol{\beta}) \, d\boldsymbol{\beta} \tag{2.2}$$

so that c is simply a normalising constant that ensures that the posterior probability distribution integrates to unity, i.e. is a correct probability density function. Given the data, then $p(\mathbf{y} \mid \boldsymbol{\beta})$ can be regarded as a function of $\boldsymbol{\beta}$ rather than of $\mathbf{y}$. In this form it is the *likelihood function* of $\boldsymbol{\beta}$ for given $\mathbf{y}$, denoted by $l(\boldsymbol{\beta} \mid \mathbf{y})$. $l(\boldsymbol{\beta} \mid \mathbf{y})$ is defined up to a multiplicative constant, so that (2.1) becomes

$$p(\boldsymbol{\beta} \mid \mathbf{y}) \propto l(\boldsymbol{\beta} \mid \mathbf{y}) p(\boldsymbol{\beta}) \tag{2.3}$$

and when the normalising constant is determined:

$$p(\boldsymbol{\beta} \mid \mathbf{y}) = c . \; l(\boldsymbol{\beta} \mid \mathbf{y}) p(\boldsymbol{\beta}) \tag{2.4}$$

The posterior probability distribution is therefore proportional to the likelihood multiplied by the prior distribution.

This Bayesian approach has a number of advantages over (or at least, contrasts with) the sampling theory approach to inference. For present purposes, four are particularly noteworthy:

1. The Bayesian approach provides a formal and easy way of incorporating prior information into the statistical analysis.
2. Bayesian procedures provide what Box and Tiao (1973) call a 'leak proof route' to inference, in that the posterior distribution makes use of all the information in the data.
3. Bayesian analysis does not rely on any asymptotic justification, and in fact Bayes' Theorem provides exact finite-sample results. This is also true of maximum likelihood estimation for many models, such as the classical linear model with normal and independent disturbances. However, in many models the only known properties of the maximum likelihood estimator are asymptotic (e.g. consistency), even when the random variables in the model are assumed normal. This is true for the maximum likehood estimator for the spatial model presented in this paper, whereas the Bayesian analysis can give finite-sample inference.
4. Bayesian procedures provide a comprehensive framework for examining the assumptions of statistical models and the robustness of models to departures from those assumptions.

7.2.1 *Prior distributions*

The specification of a prior distribution, $p(\boldsymbol{\beta})$, expressing prior knowledge about the parameters $\boldsymbol{\beta}$, is a central element in Bayesian inference. This prior distribution can express probabilities derived from earlier empirical work or from theoretical economic analysis. Such informative prior knowledge has been employed in many Bayesian econometric studies, e.g. aggregate production functions (Zellner and Richard, 1973) and simultaneous equations (Chetty, 1968). However, in many studies we wish to use the Bayesian framework without incorporating very strong prior information. This can be achieved by the employment of a 'diffuse' or 'non-informative' prior distribution. Use of such a distribution does not necessarily mean we have no prior ideas about $\boldsymbol{\beta}$, but may simply mean that we wish to use, in Box and Tiao's phrase, a 'reference prior', representing an unprejudiced state of mind, against which we can compare other assumptions (Box and Tiao, 1973).

Consider the example of the classical linear model with parameter vector $\boldsymbol{\beta}$ and normally distributed disturbances with standard deviation σ. Where one does not wish to incorporate strong prior information, the standard approach is to assume that the parameters $\boldsymbol{\beta}$ are uniformly and independently distributed:

$$p(\boldsymbol{\beta}) \propto \text{constant} \qquad -\infty < \beta_i < \infty \qquad i = 1, 2, \ldots, k$$

and where σ is also an unknown parameter, to assume that $\log \sigma$ is uniformly and independently distributed, so that the joint prior distribution is

$$p(\boldsymbol{\beta}, \sigma) \propto \frac{1}{\sigma} \qquad -\infty < \beta_i < \infty \qquad i = 1, 2, \ldots k \qquad 0 < \sigma < \infty$$

The uniform prior is strictly 'improper' since it does not integrate to unity, and has been the focus for much of the debate over Bayesian inference (see Edwards, 1972, for criticisms, and Fienberg and Zellner, 1975, for discussion). It is, however, the most common used prior in econometric work, underlying much of the analysis in the texts by Zellner (1971) and Box and Tiao (1973), and many other studies, including Zellner and Tiao's study of serial dependence in the linear regression model (Zellner and Tiao, 1965).

The use of such non-informative prior distributions leads to the dominance of the posterior distribution by the information from the likelihood function, so that the mathematical expression of the posterior distribution is similar or identical to that of the likelihood function. In particular, for many models associated with the normal distribution, the mode of the posterior distribution is identical (in terms of the $\boldsymbol{\beta}$-values) with the maximum likelihood estimates. However, the Bayesian approach presents an entire posterior probability distribution which can be interpreted, whereas the interpretation of the likelihood function at points other than its modal value is not clear, and the inferential perspective is rather different.

7.2.2 *Joint, marginal and conditional distributions*

If the parameter vector $\boldsymbol{\beta}$ has k elements, then the joint posterior distribution is a multidimensional function with a k-dimensional domain. In practice, the main focus of interest may be only one or two parameters, rather than the structure of the entire multidimensional space, and the other elements of $\boldsymbol{\beta}$ are regarded as 'nuisance parameters'. (In addition, it is extremely difficult to visualise or represent probability distributions of more than two dimensions.) In non-Bayesian inference, such nuisance parameters can only be eliminated under certain conditions by considering the conditional distribution with respect to the sufficient statistics for the nuisance parameters. In the classical linear regression model, for example, the unknown variance of the disturbances can be eliminated from the likelihood function. In contrast, however, in the maximum likelihood model for linear regression with spatially dependent errors (Hepple, 1976; Ord, 1975), one cannot eliminate the spatial autoregressive parameter as a nuisance parameter and focus on the regression coefficients.

In Bayesian analysis any nuisance parameter can be integrated out of the joint posterior distribution. The multidimensional distribution is integrated over the dimension corresponding to the nuisance parameter, and so collapsed to $k-1$ dimensions. If all parameters but one are integrated out, the resultant one-dimensional posterior distribution, known as the *marginal* posterior distribution, can be used to make 'overall' inferences about particular parameters. Consider a model with two unknown parameters, β_1 and β_2. 'Overall' inferences about β_1 can be made by integrating out β_2:

$$p(\beta_1 \mid y) = \int p(\beta_1, \beta_2 \mid y)\, d\beta_2$$

In contrast, the *conditional* posterior distribution of β_1 given β_2, $p(\beta_1 \mid \beta_2, y)$, is defined by taking a slice through the β_2 dimension at a particular value, rather than integrating over that dimension. The marginal posterior distribution of β_1 takes into account (through the integral) the probability distribution of the β_2 parameter, whereas the conditional distribution assumes β_2 is fixed at a known value.

The two distributions can easily be related. The joint posterior distribution can be expressed as the product of the conditional posterior distribution of β_1 given β_2 and the marginal posterior distribution of β_2:

$$p(\beta_1, \beta_2 \mid y) = p(\beta_1 \mid \beta_2, y)\, p(\beta_2 \mid y)$$

The marginal posterior distribution of β_1 may now be expressed as

$$p(\beta_1 \mid y) = \int p(\beta_1 \mid \beta_2, y)\, p(\beta_2 \mid y)\, d\beta_2$$

The marginal posterior distribution of the main parameter β_1 is therefore a suitably weighted function of the conditional distributions with the marginal distribution of the nuisance parameter β_2 as the weighting function.

These two forms of posterior distribution are particularly relevant to the study of the robustness of statistical models to departures from underlying assumptions. By incorporating a more general specification into the Bayesian analysis of the model (such as non-independent disturbances, non-normal disturbances, or simultaneity effects), it is possible to use the conditional posterior distributions to see how sensitive the probability distributions of the main parameters are to alternative assumptions about the 'nuisance' parameters. Where the various conditional distributions do not closely overlap, then it is clearly important to take the possible departure from standard assumptions into account and use the marginal posterior distributions that incorporate information about the probability distribution of the nuisance parameters.

When a diffuse prior is employed, the joint posterior distribution is (as noted in section 7.2.1) mathematically similar or identical to the likelihood function. This has led Maddala (1971) to comment that the important difference between maximum likelihood methods and Bayesian analysis with diffuse priors is that:

> the former amounts to a summarization of the likelihood function in terms of the location of its maximum and the curvature around the maximum, and the latter amounts to the summarization of the likelihood function in terms of some weighted averages of the likelihood. Since both give information on the behaviour of the likelihood function, which is important for any likelihood inference, Bayesian or otherwise, it is very fruitful to regard them as complementary to each other (Maddala, 1971, p. 952).

7.3 LINEAR REGRESSION MODEL WITH SPATIALLY AUTOCORRELATED DISTURBANCES

The particular model subjected to Bayesian analysis in this paper is the linear regression model with spatially autocorrelated disturbances. Consider the linear model

$$\boldsymbol{y} = \boldsymbol{X}\boldsymbol{\beta} + \boldsymbol{u} \tag{3.1}$$

where

- $\boldsymbol{y}$ is the $N \times 1$ vector of spatial observations on the dependent variable for N regions
- $\boldsymbol{X}$ is the $N \times k$ matrix of observations on the k independent variables
- $\boldsymbol{\beta}$ is the $k \times 1$ vector of regression coefficients, and
- $\boldsymbol{u}$ is the $N \times 1$ vector of disturbances.

The standard linear regression model assumes that the disturbance terms, $\boldsymbol{u}$, are independent of one another, so that the disturbance variance-covariance matrix is diagonal (and is also assumed to be homoscedastic):

$$E(\boldsymbol{u}\boldsymbol{u}') = \sigma^2 \boldsymbol{I} \tag{3.2}$$

where $\boldsymbol{I}$ is the identity matrix. In the present analysis we wish to allow for the possibility that the disturbances are spatially dependent on each other in some way, so that

$$E(\boldsymbol{u}\boldsymbol{u}') = \sigma^2 \boldsymbol{V}, \tag{3.3}$$

a non-diagonal positive definite matrix of disturbance covariances.

To proceed further the structure of $\boldsymbol{V}$ must be specified in terms of only a few unknown parameters. As in the time series case, a number of specifications are possible; here we restrict ourselves to the case where the disturbances are generated by a spatial autoregressive process of the type already analysed within a maximum likelihood framework by Ord (1975) and Hepple (1976). Alternative specifications include the spatial moving-average process, and models based directly on the elements of $\boldsymbol{V}$ rather than on the structure of the disturbance process (Lebart, 1969).

The most common form of spatial dependence is positive spatial autocorrelation between adjacent regions, so a disturbance generating process that incorporates dependence on immediate regional neighbours (as does the spatial autoregressive model) should give a parsimonious and useful representation of $\boldsymbol{V}$. The spatial autoregressive model is constructed as follows. Let $\boldsymbol{W}$ be the $N \times N$ matrix of contiguities between regions, with entry $w_{ij} = 1$ if regions i and j are contiguous, and $w_{ij} = 0$ otherwise. Now scale the row sums of $\boldsymbol{W}$ to unity, and denote this scaled matrix by $\boldsymbol{M}$. The model can now be written

$$\boldsymbol{u} = \rho \boldsymbol{M}\boldsymbol{u} + \boldsymbol{e} \tag{3.4}$$

where $\boldsymbol{e} \sim N(\boldsymbol{0}, \sigma^2 \boldsymbol{I})$. This specifies a spatial process in which each region's value is a linear function of the *average* value in contiguous regions, apart from noise $\boldsymbol{e}$. Because of the scaling of the rows in $\boldsymbol{M}$, the largest eigenvalue of $\boldsymbol{M}$ is 1.0, and this defines a theoretical upper limit to ρ of 1.0 (see equation 3.9). A similar theoretical lower limit to ρ may not exist for the regional system, but if it does it cannot be defined without knowledge of the detailed structure of $\boldsymbol{M}$ (i.e. the other eigenvalues of the matrix). However, the definition of such a theoretical lower limit is not of great practical importance. Generalization of the model to higher-order processes (e.g. second-order spatial linkages) and to generalized contact weights (allowing boundary and distance effects, as in Cliff and Ord, 1973) presents no difficulties. Denoting $\boldsymbol{P} = \boldsymbol{I} - \rho\boldsymbol{M}$, the matrix $\boldsymbol{V}^{-1}$ can be derived:

$$\boldsymbol{V}^{-1} = \boldsymbol{P}'\boldsymbol{P} \tag{3.5}$$

The full model may now be written, combining (3.1) and (3.4):

$$\begin{aligned} y &= X\beta + u \\ u &= \rho M u + e \end{aligned} \tag{3.6}$$

and this depends on a $k + 2$ vector of parameters θ, with elements

$$\{\beta_1, \beta_2, \ldots, \beta_k, \rho, \sigma\}$$

We now analyse this model in Bayesian terms, focusing in particular on the importance of the spatial autoregressive parameter, ρ, in influencing the probability distribution of β. In common with many Bayesian econometric studies, a uniform diffuse prior for θ is employed as a reference prior. An alternative approach would be to use the natural conjugate priors developed in the serial correlation case by Lempers (1971). Assuming β, ρ and $\log \sigma$ are uniformly and independently distributed gives the prior probability distribution

$$p(\beta, \rho, \sigma) \propto \frac{1}{\sigma} \qquad \begin{aligned} -\infty < \beta_i &< \infty \\ i &= 1, 2, \ldots, k \\ 0 < \sigma &< \infty \\ \rho &< 1.0 \end{aligned} \tag{3.7}$$

The analysis is not restricted to the use of a non-informative prior. On the contrary, there are circumstances in which one would wish to incorporate definite prior information. For example, many aggregate production functions are estimated using interregional cross-sectional data, and prior information can play an important role in the estimation of these models (see Zellner and Richard, 1973, for an example with time-series data).

The prior distribution (3.7) must now be combined with the likelihood function to give the posterior probability distribution for the model.

7.3.1 *Likelihood function and joint posterior distribution*

The likelihood function for the linear regression model with spatially autocorrelated disturbances of type (3.6) is

$$L \propto \frac{|P|}{\sigma^N} \exp\left\{\frac{-1}{2\sigma^2}(y - X\beta)' V^{-1} (y - X\beta)\right\} \tag{3.8}$$

A full derivation of this function is given in Hepple (1976). The quadratic exponent term is common to least-squares and linear models. $|\boldsymbol{P}|$ is the Jacobian of the transformation from the normal independent disturbances, $\boldsymbol{e}$, to the autocorrelated disturbances, $\boldsymbol{u}$, and it plays a key role in estimation (Hepple, 1976; Whittle, 1954). In the standard linear regression model (Zellner, 1971) the Jacobian is simply unity and so is eliminated from the likelihood function, and this is also true for the serial correlation model as sample size increases (Zellner and Tiao, 1965). In the spatial case $|\boldsymbol{P}|$ only collapses to unity in the rare case of a triangular structure to the weights $\boldsymbol{W}$, so that there is no simultaneity in region i influencing region j and in turn being influenced by region j. Such simultaneity is the normal case. The difference between the time-series model and the spatial model is very similar to the difference between recursive and simultaneous equation systems, and in fact $|\boldsymbol{P}|$ acts as a weighting function counteracting the tendency of the quadratic component in (3.8) to overestimate the likelihood of high ρ-values, a role very similar to that of the Jacobian of the endogenous variables in simultaneous equation models (see Chetty, 1968; Rothenberg, 1963; Zellner, 1971). $|\boldsymbol{P}|$ constrains the distribution of ρ to be less than +1.0, for $|\boldsymbol{P}|$ equals 1.0 when $\rho = 0.0$ but declines to zero at $\rho = 1.0$. For the present model $|\boldsymbol{P}|$ may usefully be simplified as follows (Ord, 1975; Hepple, 1976):

$$|\boldsymbol{P}| = \prod_{i=1}^{N} (1 - \rho \mu_i) \tag{3.9}$$

where μ_i, $i = 1, 2, \ldots, N$ are the eigenvalues of the nonsymmetric matrix $\boldsymbol{M}$. The analysis that follows is closely related to the parallel analysis for the serial correlation case, presented by Zellner and Tiao (1965), but the Jacobian $|\boldsymbol{P}|$ complicates the posterior distributions.

Combining the prior distribution (3.7) with the likelihood function (3.8) gives the posterior probability distribution

$$p(\boldsymbol{\beta}, \rho, \sigma \mid \boldsymbol{y}) \propto \frac{|\boldsymbol{P}|}{\sigma^{N+1}} \exp\left\{\frac{-1}{2\sigma^2} (\boldsymbol{y} - \boldsymbol{X}\boldsymbol{\beta})' \boldsymbol{V}^{-1} (\boldsymbol{y} - \boldsymbol{X}\boldsymbol{\beta})\right\} \tag{3.10}$$

We can now regard σ as a nuisance parameter, since it is not a primary focus of interest, and integrate it out, using the properties of the inverted gamma distribution, to give the joint posterior distribution for $\boldsymbol{\beta}$ and ρ:

$$p(\boldsymbol{\beta}, \rho \mid y) \propto |P| \{(y - X\boldsymbol{\beta})' V^{-1} (y - X\boldsymbol{\beta})\}^{-N/2} \qquad (3.11)$$

In the maximum likelihood approach to this model given in Hepple (1976), σ can also be eliminated by concentrating the likelihood to give an expression analogous to (3.11).

The joint posterior distribution in (3.11) defines a multivariate probability distribution in a multidimensional space of $k + 1$ dimensions. The Bayesian econometrician can now integrate out some of these dimensions to look at the marginal posterior distributions of particular parameters, taking into account the distributions of the omitted dimensions, or take slices through the dimensions to produce posterior distributions conditional on particular values of some of the parameters.

7.3.2 *Conditional posterior distributions*

If the spatial autoregressive coefficient ρ is fixed at some a priori value, so that ρ is no longer a random variable, we can derive the conditional posterior distribution for $\boldsymbol{\beta}$. If ρ is held fixed in expression (3.11), then $|P|$ becomes invariant across the posterior distribution and so is absorbed into the normalising constant, and the conditional posterior distribution for $\boldsymbol{\beta}$ is

$$p(\boldsymbol{\beta} \mid \rho, y) \propto \{(y - X\boldsymbol{\beta})' V^{-1} (y - X\boldsymbol{\beta})\}^{-N/2} \qquad (3.12)$$

Since the Jacobian term $|P|$ has been eliminated from the right-hand side, the form of (3.12) is directly analogous to the equivalent time-series expression, and so (3.12) can be simplified to the forms given by Zellner (1971, p. 91 and 94), i.e. to the multivariate Student t-distribution centered about the least-squares estimates of $\boldsymbol{\beta}$ for fixed ρ. If ρ is fixed at zero, then (3.12) generates the standard normal linear regression model with independent disturbances.

Note that in the conditional case, we can define first spatial differences (Martin, 1974) by setting $\rho = 1.0$, although within the framework of the spatial autoregressive model this is strictly illegitimate, since when $\rho = 1.0$ $|P|$ becomes zero, and this makes the normalising constant zero, and hence the posterior distribution.

The conditional posterior distributions for $\boldsymbol{\beta}$ reveal whether inferences about $\boldsymbol{\beta}$ are robust or sensitive to differing assumptions about ρ, i.e. to differing levels of spatial autocorrelation in the disturbances. If the conditional distributions do not closely coincide, then it is important to take the posterior distribution of ρ into account in drawing inferences about $\boldsymbol{\beta}$, by using the *marginal* distributions of the parameters.

7.3.3 *Marginal distribution for* ρ

The marginal distribution for ρ can be derived by integrating (3.11) with respect to $\boldsymbol{\beta}$:

$$p(\rho \mid y) \propto \int |\boldsymbol{P}| \{(\boldsymbol{y}-\boldsymbol{X\beta})'\boldsymbol{V}^{-1}(\boldsymbol{y}-\boldsymbol{X\beta})\}^{-N/2}\, d\boldsymbol{\beta} \tag{3.13}$$

This can be handled best by redefining the terms in the quadratic component. Recalling that $\boldsymbol{V}^{-1} = \boldsymbol{P}'\boldsymbol{P}$ and that $\boldsymbol{P} = \boldsymbol{I} - \rho\boldsymbol{M}$, then

$$\begin{aligned}&\{(\boldsymbol{y}-\boldsymbol{X\beta})'\boldsymbol{V}^{-1}(\boldsymbol{y}-\boldsymbol{X\beta})\}\\ &\quad = \{(\boldsymbol{y}-\boldsymbol{X\beta})'(\boldsymbol{I}-\rho\boldsymbol{M})'(\boldsymbol{I}-\rho\boldsymbol{M})(\boldsymbol{y}-\boldsymbol{X\beta})\}\end{aligned} \tag{3.14}$$

and defining $\boldsymbol{y}^* = \boldsymbol{y} - \rho\boldsymbol{My}$, and $\boldsymbol{X}^* = \boldsymbol{X} - \rho\boldsymbol{MX}$, we have

$$\{(\boldsymbol{y}-\boldsymbol{X\beta})'\boldsymbol{V}^{-1}(\boldsymbol{y}-\boldsymbol{X\beta})\} = \{(\boldsymbol{y}^*-\boldsymbol{X}^*\boldsymbol{\beta})'(\boldsymbol{y}^*-\boldsymbol{X}^*\boldsymbol{\beta})\} \tag{3.15}$$

After various steps integrating out $\boldsymbol{\beta}$ (see Zellner, 1971, p. 95), the marginal posterior distribution for ρ may be expressed:

$$p(\rho \mid y) \propto |\boldsymbol{P}|\ |\boldsymbol{R}|^{-1/2}\ (\boldsymbol{y}^{*\prime}\boldsymbol{G}\boldsymbol{y}^*)^{-(N-k)/2} \tag{3.16}$$

where

$\boldsymbol{G} = \{\boldsymbol{I} - \boldsymbol{X}^*(\boldsymbol{X}^{*\prime}\boldsymbol{X}^*)^{-1}\boldsymbol{X}^{*\prime}\}$, and

$\boldsymbol{R} = \boldsymbol{X}'\boldsymbol{V}^{-1}\boldsymbol{X} = \boldsymbol{X}^{*\prime}\boldsymbol{X}^*$

G is the well-known idempotent matrix that occurs in the linear regression model (Johnston, 1972, pp. 127–128), only in this instance it is in terms of X^* rather than X. Expression (3.16) differs from Zellner and Tiao's expression for the marginal posterior distribution of ρ in the time series case by the inclusion of the Jacobian term $|P|$. In contrast to the conditional distribution for $\boldsymbol{\beta}$, (3.16) does not define an analytically tractable probability distribution, and has to be computed using numerical integration techniques (see section 7.3.5).

7.3.4 *Marginal distribution for* $\boldsymbol{\beta}$

The derivation of the marginal distributions for elements of $\boldsymbol{\beta}$ requires some algebraic manipulation. In the spatial model (unlike the time-series model), the joint posterior distribution defined in (3.11) cannot be integrated analytically with respect to ρ, because of the difficulty of the $\int |P| \, d\rho$ term, and the analysis must proceed as follows. Taking expression (3.11) we integrate out all the $\boldsymbol{\beta}$-parameters *except* the one of primary interest, say β_j, leaving us with a bivariate posterior distribution in terms of β_j and ρ. This analysis follows the suggestions of Zellner and Tiao (1965). Assume that $j = 1$ for notational convenience, and partition

$$\boldsymbol{\beta}' = (\beta_1 \vdots \bar{\boldsymbol{\beta}}') \text{ and } X = (\boldsymbol{x}_1 \vdots \bar{X}). \text{ Also define}$$

$$\boldsymbol{w} = \boldsymbol{y}^* - \boldsymbol{x}_1^* \beta_1 \text{ and } \bar{X}^* = \bar{X} - \rho M \bar{X}.$$

The joint posterior distribution (3.11) may now be re-written:

$$p(\beta_1, \bar{\boldsymbol{\beta}}, \rho \mid \boldsymbol{y}) \propto |P| \, \{(\boldsymbol{w} - \bar{X}^* \bar{\boldsymbol{\beta}})'(\boldsymbol{w} - \bar{X}^* \bar{\boldsymbol{\beta}})\}^{-N/2} \tag{3.17}$$

(Note that the terms in the quadratic component, $\boldsymbol{w}$ and $\bar{X}^*$, depend on ρ). Expression (3.17) can now be integrated analytically with respect to the 'nuisance parameters' $\bar{\boldsymbol{\beta}}$ to give

$$p(\beta_1, \rho \mid \boldsymbol{y}) \propto |P| \, |\bar{R}|^{-\frac{1}{2}} \{\boldsymbol{w}' \bar{G} \boldsymbol{w}\}^{-(N-k+1)/2} \tag{3.18}$$

where

$$\bar{G} = [I - \bar{X}^*(\bar{X}^{*\prime}\bar{X}^*)^{-1}\bar{X}^{*\prime}], \text{ and}$$

$$\bar{R} = \bar{X}^{*\prime}\bar{X}^*$$

Bivariate numerical integration techniques can then be employed to determine the marginal posterior distribution for β_1. Since this procedure is necessary to obtain the marginal posterior distribution for a given $\boldsymbol{\beta}$-element, it is also convenient to compute the marginal posterior distribution for ρ from the same bivariate posterior distribution.

7.3.5 *Computational methods and numerical integration*

The practical implementation of Bayesian techniques on empirical problems depends heavily on numerical integration methods. In the present model only the conditional posterior distribution of $\underline{\beta}$ is in the form of a tractable analytical distribution. Fortunately numerical integration is a well-developed area of numerical analysis (see Dixon, 1974). The functions in many (but by no means all) Bayesian problems are numerically well-behaved and have no sudden discontinuities, and in the present case (compare also Tsurumi and Tsurumi, 1976), simple procedures such as Simpson's rule produced results as accurate as more refined techniques such as the method of Clenshaw and Curtis (1960). Simpson's rule is reviewed by Zellner (1971).

As an example of a univariate function consider the marginal posterior distribution for ρ in expression (3.16). This expression can be numerically evaluated for a range of ρ-values. The upper limit is 1.0. As noted earlier, a lower limit (LL) is more difficult to define, but because most spatial autocorrelation is positive, in practice a lower limit of -0.5 has usually been found adequate to encompass the posterior distribution. The normalising constant can then be determined by using Simpson's rule to evaluate the integral

$$c^{-1} = \int_{\text{LL}}^{1.0} f(\rho)\, d\rho \tag{3.19}$$

where $f(\rho)$ represents the value of (3.16) for a specific value of ρ.

The function can now be normalised to define a proper density function that integrates to unity:

$$p(\rho) = c\,f(\rho) \tag{3.20}$$

This marginal posterior probability density function can either be represented graphically, or summarised in terms of its first four moments.

For bivariate integration of expression (3.18) one evaluates (3.18) numerically for a whole grid of pairs of β_1 and ρ values. The normalising constant is then determined by the double application of Simpson's rule:

$$c^{-1} = \int_{a}^{b} \int_{\mathrm{LL}}^{+1.0} p(\beta_1, \rho)\, d\rho \quad d\beta_1 \tag{3.21}$$

where a and b are the lower and upper limits for β_1. The marginal distributions for β_1 and ρ can then be evaluated by integrating over either the rows or columns of the normalised grid of values. The accuracy of the integrations and hence of the density function can be improved by increasing the number of steps (intervals) in the numerical evaluation and integration. (This means that the *exactness* of Bayesian finite-sample inference is dependent on the computational accuracy of the integrations.)

7.4 AN EMPIRICAL APPLICATION

To illustrate the application of the Bayesian techniques developed in previous sections, the linear regression model given in Hepple (1976) was subjected to Bayesian analysis. The data derive from a study of the effect of sales taxes and transport charges for new cars on the second-hand value of cars in the 49 states of the U.S.A. The regression equation is

$$\boldsymbol{y} = \beta_1 + \beta_2 \boldsymbol{x} + \boldsymbol{u} \tag{4.1}$$

where $\boldsymbol{y}$ is the vector of 49 observations on the average value of 1955–1959 model cars in January 1960 for each state, and $\boldsymbol{x}$ is the vector of observations on new-car price differentials attributable to transport charges and sales taxes. Both variables are expressed in dollars. β_2 is clearly expected to be >0 but probably <1.

Ordinary least-squares gives a $\hat{\beta}_2$ estimate of 0.686, with a t-value of 3.52. However, the regression residuals show strong spatial autocorrelation, and Hepple (1976) re-estimated the equation using the maximum likelihood model incorporating spatial autoregression in the disturbances. The maximum likelihood estimate of ρ is 0.816, and that for β_2 is now 0.099, which is no longer significantly different from zero.

For the Bayesian analysis (as for the maximum likelihood analysis), the $\boldsymbol{W}$-matrix was constructed from the binary contiguities between states, with the weights then scaled to give a matrix $\boldsymbol{M}$ with row sums equal to one. The eigenvalues of $\boldsymbol{M}$ (used in the calculation of the Jacobian $|\boldsymbol{P}|$) where computed using the procedures in Wilkinson and Reinsch (1971).

For the Bayesian analysis of the model a diffuse prior distribution was used:

$$p(\beta_1, \beta_2, \rho, \sigma) \propto \frac{1}{\sigma} \tag{4.2}$$

Combining (4.2) with the likelihood function given in (3.8) produces the joint posterior distribution in (3.10) and integrating out σ gives (3.11). The intercept term was also integrated out as a nuisance parameter, and attention focused on β_2 and ρ.

To examine the sensitivity of the posterior distribution of β_2 to assumptions about the value of ρ, conditional posterior distributions for β_2 were computed and reported in Fig. 1. The conditional distributions reveal that the distribution of β_2 is the fact sensitive to the assumed value of ρ. There is, for example, only limited overlap between the distributions for $\rho = 0.0$ and $\rho = 0.75$. The modal (and mean) value of β_2 for $\rho = 0.0$ is the OLS estimate of 0.686, but if ρ is really 0.75, then a β_2 value of 0.686 is remote, i.e. more than four standard deviations from the mean value, which is now 0.138. If true ρ is greater than 0.75, the divergence is even more marked. Since ρ is in fact unknown it is therefore important to make inferences about β_2 taking into account the posterior distribution of ρ, i.e. to use the marginal posterior distribution β_2.

The bivariate posterior distribution for β_2 and ρ is given in Fig. 2. This is constructed from a grid of 2500 values. The maximum point

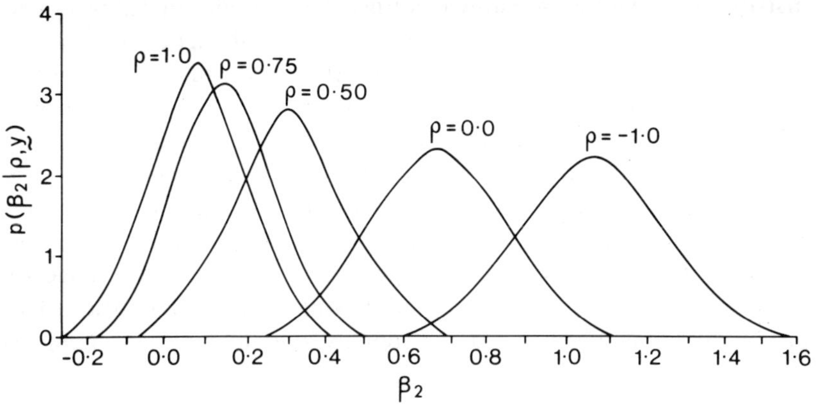

Fig. 1 Conditional posterior distributions for β_2

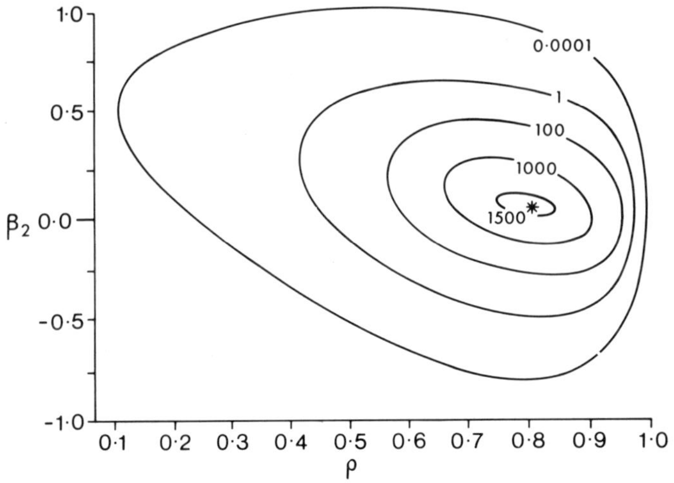

Fig. 2. Bivariate posterior distribution for β_2 and ρ

corresponds to the maximum likelihood estimates of β_2 and ρ (the grid evaluations do not locate the maximum itself to the same degree of accuracy as the maximum likelihood search; the highest grid point is in fact $\beta_2 = 0.10$ and $\rho = 0.81$). Bivariate integration using Simpson's rule was employed to calculate the marginal posterior distributions. These are shown in Figs. 3 and 4, and in Table 1. The marginal posterior distribution for β_2 shows a strong concentration around 0.00 – 0.20, with the distribution being highly symmetric about the mode and with a kurtosis close to that for the conditional distributions. The marginal distribution reveals that a β_2 value of 0.686 is very unlikely when the probability distribution of ρ is taken into account. It is now useful to examine the marginal posterior distribution for ρ, which is concentrated close to the mode of 0.81 and is slightly negatively skewed with a range of values between 0.70 and 0.95 all having high probabilities. The probability that ρ is near 0.0 is extremely low; in fact the probability of any value below 0.5 is negligible.

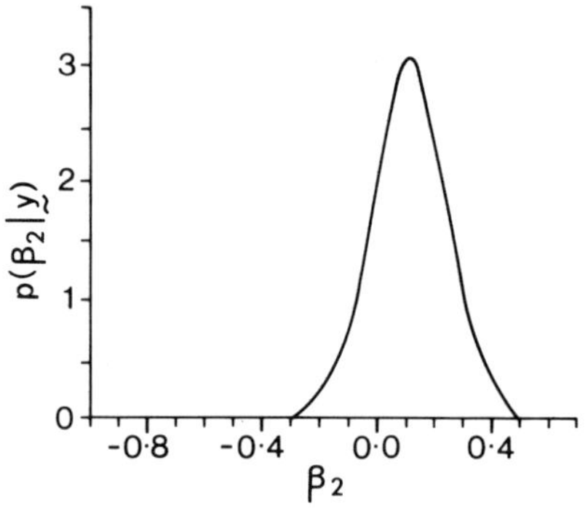

Fig. 3. Marginal posterior distribution for β_2

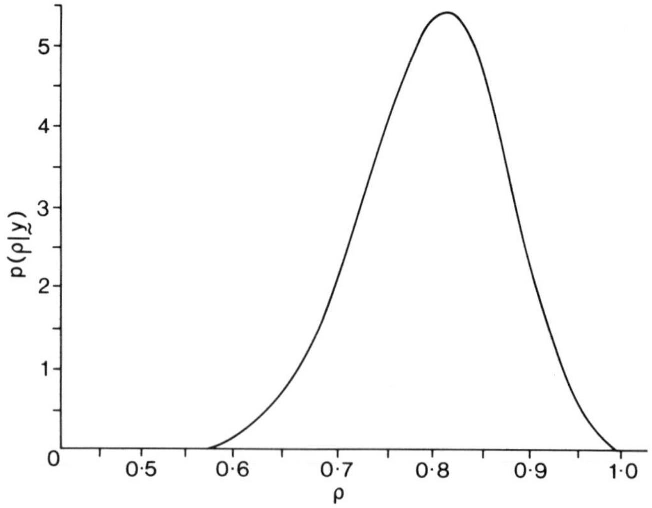

Fig. 4. Marginal posterior distribution for ρ

Table 1. Summary measures of marginal distributions

	β_2	ρ
Mode	0.10	0.81
Mean	0.104	0.801
St. Dev.	0.131	0.074
Skewness	0.049	-0.498
Kurtosis	3.231	3.333

These results show the usefulness of the Bayesian approach in examining the robustness or sensitivity of a linear regression model to spatial dependence in the disturbances. The marginal probability distribution of β_2 is markedly different from the original posterior distribution based on the standard normal regression model (equivalent to the conditional distribution with $\rho = 0.0$). Similarly the marginal distribution of ρ has shown that zero or negative spatial dependence is extremely improbable for this model. The approach has two advantages over the maximum likelihood analysis of the same problem: the lack of any appeal to asymptotic justification, and the study of the full shape of the posterior distribution rather than just the maximum and local curvature of the likelihood function. However, like Maddala (1971), we would emphasise the complementarity rather than the rivalry of the two approaches.

7.5 CONCLUDING REMARKS

The object of this paper has been to begin the building of a bridge between recent developments in spatial statistics and spatial econometrics and the large and influential body of Bayesian work in mainstream econometrics. There has only been space here to explore the Bayesian analysis of one specific model, the linear regression model with spatially autoregressive disturbances, and many other models and procedures relevant to spatial econometrics can be profitably studied by Bayesian techniques, such as the Bayesian comparison of alternative models (Gaver and Geisel, 1974), a theme that will be taken up in another paper.The present analysis of the linear model with spatial dependence has, however, shown how Bayesian perspectives can provide alternative (and complementary) viewpoints to those provided by maximum likelihood analysis. Even if one does not place emphasis on the finite-sample inference of Bayesian methods (an emphasis the present writer *would* want to make), then Bayesian approaches provide a fruitful alternative way of examining the likelihood function. Of course, some of the major merits of the Bayesian approach are also the cause of its main limitation. The full exploration of the shape of the posterior distribution makes the Bayesian approach computationally unwieldy for problems with large numbers of parameters that do not simplify easily. If the present analysis was applied to a model with a large number of independent variables, then the computational burden would soon become great. This burden has in practice limited the usefulness of Bayesian analysis for most simultaneous equation systems in econometrics. Where it is not feasible to explore the entire shape of the posterior distribution, it is often still computationally easy to find a route to the maximum of the likelihood function, even for an enormous number of parameters. However, it is a less rich analysis, and for smaller models the Bayesian provides a very valuable tool.

REFERENCES

Box, G. E. P. and Tiao, G. C. (1962), 'A further look at robustness via Bayes Theorem', *Biometrika*, 49, 419–433.

Box, G. E. P. and Tiao, G. C. (1973), *Bayesian inference in statistical analysis*, Addison-Wesley, Reading, Massachussetts.

Chetty, V. K. (1968), 'Bayesian analysis of Haavelmo's models', *Econometrica*, 36, 582–602.

Clenshaw, C. W. and Curtis, A. R. (1960), 'A method for numerical integration on an automatic computer', *Numeriche Mathematik*, 2, 197–205.

Cliff, A. D. and Ord, J. K. (1973), *Spatial autocorrelation*, Pion, London.

Dixon, V. A. (1974), 'Numerical quadrature: a survey of the available algorithms', in D. J. Evans (ed.), *Software for numerical mathematics*, Academic Press, London, 105–137.

Edwards, A. W. F. (1972), *Likelihood*, Cambridge University Press, Cambridge.

Fienberg, S. E. and Zellner, A. (eds.) (1975), *Studies in Bayesian econometrics and statistics*, North Holland, Amsterdam.

Gaver, K. M. and Geisel, M. S. (1974), 'Discriminating among alternative models: Bayesian and non-Bayesian methods', in P. Zarembka (ed.), *Frontiers in Econometrics*, Academic Press, New York, 49–77.

Hepple, L. W. (1976), 'A maximum likelihood model for econometric estimation with spatial series', in I. Masser (ed.), *Theory and practice in regional science*, Pion, London, 90–104.

Johnston, J. (1972), *Econometric methods*, (2nd ed.), McGraw-Hill, New York.

Lebart, L. (1969), 'Analyse statistique de la contiguite', *Publications de l'Institut de Statistique de l'Universite de Paris*, 18, 81–112.

Lempers, F. B. (1971), *Posterior probabilities of alternative linear models*, Rotterdam University Press, Rotterdam.

Maddala, G. S. (1971), 'The likelihood approach to pooling cross-section and time-series data', *Econometrica*, 39, 939–953.

Martin, R. L. (1974), 'On spatial dependence, bias and the use of first spatial differences in regression analysis', *Area*, 6, 185–194.

Ord, J. K. (1975), 'Estimation methods for models of spatial interaction', *Journal of the American Statistical Association*, 70, 120–126.

Rothenberg, T. J. (1963), 'A Bayesian analysis of simultaneous equation systems', *Report 6315, Econometric Institute, Netherlands School of Economics*, Rotterdam.

Tsurumi, H. and Tsurumi, Y. (1976), 'A Bayesian estimation of macro and micro CES production functions', *Journal of Econometrics*, 4, 1–25.

Whittle, P. (1954), 'On stationary processes in the plane', *Biometrika*, 41, 434–449.

Wilkinson, J. H. and Reinsch, C. (eds.) (1971), *Handbook for automatic computation: Vol. II, linear algebra*, Springer-Verlag, Berlin and Heidelberg.

Zellner, A. (1971), *An introduction to Bayesian inference in econometrics*, Wiley, New York.

Zellner, A. and Geisel, M. S. (1970), 'Analysis of distributed lag models with applications to consumption function estimation', *Econometrica*, 38, 865–888.

Zellner, A. and Richard, J. F. (1973), 'Use of prior information in the analysis and estimation of Cobb-Douglas production function models', *International Economic Review*, 14, 107–119.

Zellner, A. and Tiao, G. C. (1965), 'Bayesian analysis of the regression model with autocorrelated errors', *Journal of the American Statistical Association*, 59, 763–778.

8. Alternative methods of estimating spatial interaction models and their performance in short-term forecasting

STAN OPENSHAW*

8.1 INTRODUCTION

The theoretical basis for the family of spatial interaction models and their generation using mathematical model building techniques are well known (Wilson, 1967, 1970). Indeed models based on these principles are widely used in applied studies to provide a macro-representation of cross-sectional patterns of spatial interaction. Despite enduring criticisms that they lack a satisfactory micro-behavioural rationale (Sayer, 1976; Gough, 1976), it is far from obvious what should or can be used to replace existing models, given the severe data constraints that characterise applied study. So there are good pragmatic reasons for investigating further the properties of spatial interaction models as an aggregate statistical representation of an apparently widely recurrent empirical regularity.

During the last decade most research in this field has been focused on a largely theoretical model building approach. On reflection, it appears that many researchers have been blinkered by the implicit criterion of validity that has become associated with an entropy-maximising approach. However, the applied use of spatial interaction models is dependent on the model being able to reproduce an observed situation as well as being able to generate forecasts of 'reasonable' accuracy for 10 or more years into the future. In this context, it is suggested that the advantages offered by formal model building techniques should be assessed by the benefits they bring to model performance rather than by the degree of mathematical consistency they offer. Unfortunately, entropy-maximising

*Department of Town and Country Planning, Newcastle University, Newcastle upon Tyne, NE1 7RU, England.

methods have brought little or no improvement in model performance, nor was this their intention.

In the usual model building paradigm the calibration stage is responsible for the specification of a model, the selection of a deterrence function, the estimation of the undetermined parameters, and an assessment of model goodness-of-fit in terms of its ability to reproduce base year data. If the model performs poorly in this most favourable situation then it should not be used. Rather surprisingly few studies have been explicitly concerned with model goodness-of-fit during calibration (Hathaway, 1975; Black and Salter, 1975; Openshaw, 1975, 1976). Furthermore, the model validation stage is usually incomplete since all the available data are used for calibration.

Validation includes testing the specification of the model using data for different study regions and tries to answer the question, 'Does the model generally provide an acceptable description of spatial interaction data?' It also involves testing a specific model and the values assigned to the parameters of the model for a single study region using data based on either different samples drawn from the same population at a given moment of time, or on samples drawn from possibly different populations for two or more points in time. The latter is important because it is closely related to testing the forecasting ability of a model. Once again it is surprising that so little attention has been given to this form of validation because of the widespread use of spatial interaction models in planning as forecasting tools.. Possible exceptions include Byler and O'Sullivan (1974), and Snickars and Weibull (1977).

This paper is concerned with a retrospective empirical study of the performance of a selection of singly and doubly constrained spatial interaction models in terms of their ability to forecast observed data for the same study region at two points in time. The characteristics of the data and the models are described in section 8.2. The paper is particularly concerned with a comparison of the relative efficiency, in terms of goodness-of-fit to observed data, of alternative methods of estimating what is basically the same model. This involves a consideration of both statistical and spatial aspects. The statistical content relates to the effects of alternative methods of parameter estimation on model performance as well as a consideration of data errors (see sections 8.3 and 8.4). Spatial aspects assume a greater significance in section 8.5 when the specification of the model is changed to permit a family of deterrence functions each with an explicit spatial domain. Brief consideration is also given to the problems which result from the non-uniqueness of spatially aggregated spatial interaction data. Finally, section 8.6 considers some of the implications of the results for the use of spatial interaction models in applied study.

8.2 DESCRIPTION OF DATA AND MODELS

8.2.1 *Data*

A study of either forecast performance or a test of model validity requires that observed spatial interaction data are available for the same zoning system for at least two time periods. Perhaps the major reason for the virtual absence of such studies are the tremendous data problems that this involves. For example, each new transportation study usually starts with a new zoning system, and even when a number of household surveys provide interaction data for the same region the data are often coded to different and incompatible zoning systems.

In the UK, national journey to work information for a ten per cent sample of households is available at census enumeration district level for 1966 and 1971, the latter only becoming available in 1975–6. Unfortunately, the definition of enumeration districts changed between the censuses, with the result that few zoning systems defined for use with 1966 census data can be accurately redefined for the 1971 census if the zones are not based on either ward of Local Authority sized building blocks. More seriously, it is virtually impossible to re-create the travel cost networks which characterised the historic 1966 and 1971 situations, unless this information were collected at the time and preserved.

The data used in this study relates to a region in North-East England based on Durham County and is covered by 73 zones, 10 of which are external zones representing the neighbouring areas of Tyneside and Cleveland. This study region is probably unique in that historically accurate network times are available for both 1966 and 1971. Furthermore, the size of the zones are such that they can be completely defined in terms of 1966 and 1971 census data. However, the network times accurately represent only car based trips, so attention is restricted to this data, thereby avoiding the need to allow for parameter changes in some generalised travel-cost function. Table 1 summarises the main features of this data, the trip component of which relates to the ten percent sample of all trips, provided by the censuses. It is interesting to note that between 1966 and 1971 employment levels declined slightly but the number of car based journeys to work increased tremendously. Furthermore, between 1966 and 1971 a major motorway network had been built resulting in large reductions in travel times and a decline in mean trip length. Thus this data should provide a good test of the spatial interaction model's ability to generate short-term forecasts of cross-sectional data in a non-static situation.

Table 1. Trip data characteristics for the study region in 1966 and 1971

	1966	1971
Number of car trips (10 per cent sample)	6,955	14,213
Total trips all modes (10 per cent sample)	36,704	35,254
Mean trip time in minutes	13.8	11.7
Percentage of intra-zonal trips	22.7	32.2

8.2.2 *Definition of models*

Attention is restricted to basic singly and doubly constrained models which are non-linear in their parameters because most planning applications involve models of this form. Sometimes these basic models are embedded within a larger systems modelling framework, for example, Lowry models. However, this paper concentrates on the basic spatial interaction model for a homogenous person-type category in order to examine the performance of the model, free from various embellishments designed to improve its theoretical, vis-a-vis empirical properties.

The principal model used in this study is singly constrained and it can be expressed as a function of its undetermined parameters:

$$T_{ij}(\boldsymbol{\varphi}) = O_i D_j B_j(\boldsymbol{\varphi})\, \beta(\boldsymbol{\varphi}: c_{ij}) \tag{2.1}$$

where

$$B_j(\boldsymbol{\varphi}) = \left(\sum_i O_i\, \beta(\boldsymbol{\varphi}: c_{ij})\right)^{-1} \text{ for all } (i, j) \in G, \tag{2.2}$$

T_{ij} is the predicted number of trips from origin zone i to destination zone j

O_i	is the number of trips starting from origin zone i
D_j	is the number of trips ending in zone j
$\beta(\varphi : c_{ij})$	is a function of the travel cost variable c_{ij} and is characterised by a set of unknown parameters; $\{\varphi_i\}$ $i = 1, 2, \ldots n$
c_{ij}	is a measure of the time taken to travel by car from zone i to zone j, and
G	is the set of all (i, j) contained in the study region.

A doubly constrained model is also examined. This is of the form

$$T_{ij}(\varphi) = O_i D_j A_i(\varphi), B_j) B_j(\varphi, A_i) \beta(\varphi : c_{ij}) \tag{2.3}$$

where

$$A_i(\varphi, B_j) = (\sum_k B_k(\varphi, A_i) D_k \beta(\varphi : c_{ik}))^{-1}, \tag{2.4}$$

$$B_j(\varphi, A_i) = (\sum_k A_k(\varphi, B_j) O_k \beta(\varphi : c_{kj}))^{-1} \text{ for all } (i, j) \in G \tag{2.5}$$

Equation (2.2) ensures that the model's predictions are constrained at the destination end, thus

$$\sum_i T_{ij}(\varphi) = D_j \tag{2.6}$$

while equations (2.4) and (2.5) ensure that the doubly constrained model in equation (2.3) is also constrained at the origin end, thus

$$\sum_j T_{ij}(\varphi) = O_i \tag{2.7}$$

It is noted that it makes no difference whether these models are obtained by analogy with Newton's gravity law or by entropy-maximising methods. Furthermore, in equations (2.1) and (2.3) both models are given a deterministic representation; in fact, they require the introduction of a stochastic disturbance term before attempts can be made to obtain statistical estimates of the parameter set $\boldsymbol{\varphi}$.

8.3 PARAMETER ESTIMATION AND MODEL CALIBRATION IN TERMS OF 1966 AND 1971 DATA

8.3.1 *Parameter estimation or model calibration?*

This section is concerned with investigating the effects on model performance and parameter values of different methods of parameter estimation in both 1966 and 1971. The conventional econometric approach is to choose a parameter estimation method which is thought likely to provide the most accurate and least biased sample estimates of the undetermined parameters $\boldsymbol{\varphi}$. However, in many instances the most important consideration is the implicit assumption that the parameters themselves are of primary interest, and that their accurate estimation is important to the use of the model in both calibration year and forecasting exercises. This point of view can be contrasted with one which sees parameter estimation as a means of obtaining a best fit model in terms of the data. Obviously a model should provide an acceptable fit to the data which it is supposed to represent, but acceptability is difficult to define and unlike the parameters precise tests cannot usually be formulated. Nevertheless, the planner using a model will find little comfort in the fact that the model's parameters are unbiased and significantly different from zero if the model's fit is judged to be poor.

Ideally, if a model is correctly specified then the definition of unbiased parameters will also ensure a best fit model. Unfortunately, in practice, little confidence can be expressed in the accuracy of the model's specification. Furthermore, the choice between parameter estimation and model calibration is even more important for models which are non-linear in their parameters, because methods which may provide the best parameter estimates often do so at the expense of model fit whereas methods which optimise fit may do so at the expense of parameter accuracy (Stetzer, 1976).

The choice between parameter estimation and model calibration is not clear-cut since strong technical arguments can be made for both points of view. Moreover, it is usually impossible to predict what the most appropriate method is likely to be. It is useful, therefore, to consider a number of different estimators and measure their ability to reproduce 1966 and 1971 interaction data. For this purpose attention is focused on a negative exponential deterrence function for both equations (2.1) and (2.3), that is

$$\beta(\boldsymbol{\varphi}:c_{ij}) = \exp(-\varphi_1 c_{ij}) \tag{3.1}$$

Finally, it is noted that no attempts are made to take into account specification errors in the models, or to provide tests of parameter significance because the special nature of the interaction data means that conventional tests would almost certainly be incorrectly applied.

8.3.2 *Maximum-likelihood estimation (MLE)*

The most commonly used method of parameter estimation is the method of maximum likelihood (Evans, 1971; Batty and Mackie, 1972). Assuming the observed trips have a multinomial distribution, then a maximum likelihood estimation of φ_1 can be obtained by satisfying the first-order condition for a maximum

$$\sum_i \sum_j T_{ij}(\varphi_1)c_{ij} - \sum_i \sum_j N_{ij} c_{ij} = 0 \tag{3.2}$$

where N_{ij} is the observed number of trips from zone i to zone j. The constraint can be solved numerically, for example, by minimising the quadratic function $F(\varphi_1)$:

$$F(\varphi_1) = \left(\sum_i \sum_j T_{ij}(\varphi_1) c_{ij} - \sum_i \sum_j N_{ij} c_{ij}\right)^2 \tag{3.3}$$

This constraint also occurs in the entropy-maximising derivation of the model, as a macro-state travel cost constraint. Indeed, unless a value is choosen for φ_1 which satisfies the constraint, the model cannot be regarded as an entropy-maximising one. It is also noted that a value for φ_1 which satisfies this constraint is completely dominated by the observed part of the equation, regardless of the fit of the model to the data.

8.3.3 *Nonlinear least-squares (NLLS)*

The method of maximum-likelihood is appropriate only when the form of the sampling distribution of the observed trips is known. In practice relatively little is known about the nature of this particular sampling distribution, so it is necessary to consider other methods of estimation that do not necessarily require a distributional assumption. These methods are based on different measures of the divergence between the observed and predicted trips.

The method of least-squares is perhaps the most widely used curve-fitting technique, due to its mathematical and computational simplicity. It involves minimising the sum of the squared differences between the predicted and observed trips, that is, minimise $F(\varphi_1)$:

$$F(\varphi_1) = \sum_i \sum_j (T_{ij}(\varphi_1) - N_{ij})^2 \tag{3.4}$$

This method can also be used in hypothesis testing provided the model has an additive error term which is normally distributed, homoscedastic, and uncorrelated. In the case of the spatial interaction model the assumptions of homoscedasticity and zero spatial autocorrelation of the residuals are unlikely to be satisfied (Openshaw, 1976).

8.3.4 *Minimum sum of absolute errors (MSAE)*

In certain problems minimising the sum of absolute errors, or L_1 norm may be more appropriate than NLLS. L_1 norm estimation is more robust to outliers and to heavy-tailed error distributions; for example, double exponential and various contaminated normal distributions (Rice and Young, 1964). This involves minimising $F(\varphi_1)$:

$$F(\varphi_1) = \sum_i \sum_j \left| T_{ij}(\varphi_1) - N_{ij} \right| \tag{3.5}$$

This function is difficult to minimise since first order derivatives cannot be obtained analytically. One solution is to use a numerical approximation of the derivatives, another is the iterative weighting algorithm of Schlossmacher (1973).

8.3.5 *Minimax (MMX)*

Another approach is to use a minimax or L_∞ norm estimator. This method is very conservative and merely seeks to minimise the absolute value of the maximum residual; minimise $F(\varphi_1)$,

$$F(\varphi_1) = \max_{(i,j)\in G} \left| T_{ij}(\varphi_1) - N_{ij} \right| \tag{3.6}$$

This function is also difficult to minimise but an efficient method has been developed by Madsen (1973).

8.3.6 *Minimum chi-square (MCS)*

A minimum chi-square estimator may be appropriate when the observed trip pattern is subject to sampling error. It provides a test of the hypothesis that both the predicted and observed distributions come from the same population. However, published chi-square tables cannot be used to test this hypothesis because of the special features of trip matrices, especially their large size and high proportion of cells with zero observations and low model frequencies (Leese, 1977). Nevertheless, minimising a chi-square statistic is still equivalent to optimising the fit of the model, so minimise $F(\varphi_1)$:

$$F(\varphi_1) = \sum_i \sum_j (T_{ij}(\varphi_1) - N_{ij})^2 \,/\, T_{ij}(\varphi_1) \tag{3.7}$$

This estimator presents a number of practical difficulties. It is best regarded as a weighted sum-of-squares function which gives greater emphasis to relative rather than absolute errors and no longer assumes that the sampling errors are homoscedastic.

8.3.7 *Two more weighted least-squares estimators (WLS1, WLS2)*

A number of potentially useful weighted least-squares estimators can also be used. The chi-square estimator can be modified to provide a percentage error estimator (WLS1), so minimise $F(\varphi_1)$:

$$F(\varphi_1) = \sum_i \sum_j (T_{ij}(\varphi_1) - N_{ij}) \,/\, T_{ij}(\varphi_1)^{\,2} \tag{3.8}$$

Alternatively, it may be thought desirable to weight the sum-of-squares function by the inverse of the associated sampling variances for the observed trips (WLS2), so minimise $F(\varphi_1)$:

$$F(\varphi_1) = \sum_i \sum_j (T_{ij}(\varphi_1) - N_{ij})^2 \,/\, w_{ij} \tag{3.9}$$

where w_{ij} is the variance of the sampling error associated with N_{ij}. These weights may also be generated in a number of other ways but the main problem is to ensure that the function $F(\varphi_1)$ remains convex in the domain of interest.

8.3.8 *Function minimisation*

In all cases the identification of an optimal value for φ_1 involves finding the minimum of functions which are non-linear with respect to this parameter. Thus unconstrained optimisation procedures have to be used and these will readily identify only a local optimum and not necessarily the global solution. Moreover, considerable care is needed because the functions may not be defined for all possible parameter values and may not be globally convex. In this study analytical first derivatives are provided for the MLE, NLLS, MCS, WLS1, and WLS2 estimators. These functions are minimised using Harwell subroutines VDO2AD (Hopper, 1973). The MMX function is minimised using VGO1AD. Numerical derivative approximations are used to minimise the MSAE estimator in subroutine VDO1AD. Equivalent or similar routines are available in the NAG subroutine library. Convergence in all cases is measured in terms of the absolute change in the parameter values between iterations being less than 10^{-8}, to ensure comparability between different methods.

8.3.9 *Measuring goodness-of-fit*

Once estimates have been made of φ_1, it is necessary to consider ways of assessing how well the predicted $T_{ij}(\varphi_1)$ values correspond to the observed N_{ij} values. In fact many of the criteria suggested as estimators can also be used as measures of model goodness-of-fit. Indeed the same problems arise concerning the choice of a metric in which to measure the divergence between the observed and predicted values of the dependent variable. Previous experience has shown that it does not matter which statistic is used for interpretation (Openshaw, 1976). The major problem concerns translating it into a meaningful statement about model performance and then deciding what represents an acceptable lower limit. A useful statistic is the residual standard deviation, $\hat{\sigma}$

$$\hat{\sigma} = \left(\sum_i \sum_j (T_{ij}(\varphi_1) - N_{ij})^2 / p\right)^{1/2}, \tag{3.10}$$

where p is the number of trip pairs in G which in this instance amounts to 5229. Clearly this statistic is closely related to NLLS estimation and is at a minimum when this method is used. Moreover, its magnitude reflects the size of the trips in the study region; consequently, it cannot be used to compare model performances in different years. Nevertheless it is a sensitive measure of model performance and makes possible a comparison of alternative estimation procedures in relation to NLLS. Since $\hat{\sigma}$ is related to the variance of the error term, it is possible to compare its magnitude with the trip size distribution of the observed trips as a means of obtaining an indication of what level of error is acceptable. Obviously it is desirable to try and ensure that as many trips as possible exceed $\hat{\sigma}$ in magnitude. Thus the relative performance of the model can be measured by the percentage of non-zero observed trips which exceed one residual standard deviation – the higher this statistic the better the performance of the model.

8.3.10 *Parameter estimates and model goodness-of-fit in 1966 and 1971*

Values are assigned to the undetermined parameter φ_1 in both models using seven different estimators for the 1966 and 1971 data (see table 2). Since there is only one parameter to be estimated, comparison of the results is relatively straightforward. A wide range of parameter values is hardly surprising bearing in mind the different nature of the various estimators. The question now arises as to whether or not these differences are likely to be statistically significant. A precise test requires various assumptions to be satisfied and this is unlikely to be achieved with these models. However, an indication of the possible significance of these differences in φ_1

Table 2. Parameter values and model goodness-of-fit in 1966 and 1971

	Singly constrained model				Doubly constrained model			
	1966		1971		1966		1971	
Estimator	φ_1	$\hat{\sigma}$	φ_1	$\hat{\sigma}$	φ_1	$\hat{\sigma}$	φ_1	$\hat{\sigma}$
MLE	.128	3.5	.139	9.5	.142	2.4	.168	4.7
NLLS	.130	3.5	.104	9.1	.160	2.3	.211	4.4
MSAE	.159	3.7	.180	10.1	.179	2.3	.195	4.4
MMX	.127	3.5	.074	9.8	.173	2.3	.246	4.5
MCS	.073	4.4	.084	9.4	.082	3.7	.101	7.2
WLS1	.043	5.4	.054	11.4	.050	4.9	.062	10.5
WLS2	.119	3.6	.237	11.2	.141	2.4	.244	4.5

can be obtained by calculating the standard errors of the NLLS parameters (see Bard, 1974). In this instance the standard errors for both models and both data sets are approximately 0.002. The results show that the changes in φ_1 between 1966 and 1971 far exceed these standard errors, so do most of the differences between the various estimators within each time period.

Except for the MCS and WLS1 estimators, the range in parameter values is considerably larger than the associated changes in $\hat{\sigma}$. This would suggest that both models are relatively insensitive to quite large proportional changes in their parameters. Thus the choice of estimator is largely a matter of personal preference with none being outstanding either in terms of model goodness-of-fit or stability over time. Furthermore, there is evidence that model performance has actually declined in relative terms between 1966 and 1971. The percentage of trips which exceed one residual standard deviation for the singly constrained model range from 25 to 31 per cent in 1966, and from 14 to 18 per cent in 1971. Equivalent figures for the doubly constrained model are 25 to 56 per cent in 1966 and 14 to 35 per cent in 1971. In both years there must be strong doubts about whether or not the fit of the models is sufficient for a reasonable description of the data.

8.4 ON THE ACCURACY OF SHORT-TERM FORECASTS MADE BY SPATIAL INTERACTION MODELS

8.4.1 *The use of static models in forecasting*

Spatial interaction models are widely used to forecast future patterns of interaction. Yet it is well known that there is a major logical inconsistency in the use of any static model in a dynamic context, since the model has been calibrated to reflect a particular year, whereas it is being used to forecast marginal change in another year (Batty, 1976, pp. 92-3). Nevertheless, static models are used for forecasting purposes because existing data series are generally insufficient to permit the development of more dynamic representations. Thus the important question arises as to whether or not static spatial interaction models are capable of generating even short term forecasts which possess any degree of accuracy. A related matter concerns the relative efficiency of different methods of parameter estimation. These and other questions can only be answered by an ex post facto study of model performance in a forecasting context.

8.4.2 *Forecasting and back-forecasting 1966 and 1971 trip patterns*
In a conventional planning study, exogenous forecasts would be made for the O_i, D_j, and c_{ij} variables for some future date and the model used to generate forecasts of the associated trip distribution patterns. The parameter φ_1 would be held constant at its calibration year value. However, in this study forecasts are made for 1971 using observed 1971 values for the O_i, D_j, and c_{ij} variables so that the only source of error results from the use of possibly inappropriate 1966 parameter values. This exercise can be repeated by back-forecasting the 1966 pattern of trips using 1971 parameter values with observed 1966 values for O_i, D_j, and c_{ij} (see table 3).

Table 3. Goodness-of-fit of 1966 forecasts of 1971, and 1971 forecasts of 1966

Estimators	Singly constrained model 1966 forecast of 1971, $\hat{\sigma}$	Singly constrained model 1971 forecast of 1966, $\hat{\sigma}$	Doubly constrained model 1966 forecast of 1971, $\hat{\sigma}$	Doubly constrained model 1971 forecast of 1966, $\hat{\sigma}$
MLE	9.3	3.5	4.9	2.3
NLLS	9.3	3.7	5.4	2.6
MSAE	9.8	4.0	4.6	2.5
MMX	9.3	4.4	4.7	2.9
MCS	9.9	4.1	8.5	3.1
WLS1	13.1	5.0	12.1	4.4
WLS2	9.4	4.8	5.4	2.9

A comparison of tables 2 and 3 suggests that the forecast performance of both models are about the same as their calibration year performance. This was not expected. Indeed some of the forecasts are actually better than the equivalent calibration year performances when measured in terms of $\hat{\sigma}$. Thus despite quite large changes in parameter values between 1966 and 1971 the effects of using 1966 parameters to describe the 1971 pattern of trips is not much worse than their use to describe the pattern of trips in 1966. These results are obtained despite very considerable changes in the pattern of trips between 1966 and 1971; for example, a New Town was built during this period. It is by no means a static situation either in terms of the distribution of population or employment, yet the assumption of static parameters has a none too serious effect on the forecast performance of this conventional spatial interaction model.

8.4.3 *On the significance of forecast errors in the trip end totals*
In any real planning study the O_i, D_j, and c_{ij} variables are not

accurately known for the forecast year. At best they can be regarded as mid-points of a distribution of values; at worst, confidence limits cannot be provided. Indeed forecasts of the O_i and D_j variables would normally be produced for only a small number of large zones, for example, Local Authorities and Employment Exchanges. These forecasts would then be spatially disaggregated to provide estimates for each of the model zones. Typically, confidence limits for these forecasts are large even before zonal disaggregation; for example, 15 year Box-Jenkins employment forecasts based on a 22 year series can easily have 90 per cent confidence limits of plus or minus 50 to 90 per cent of the 1971 levels. To these forecasting uncertainties must be added the unknown additional errors of disaggregation. It is useful, therefore, to assess the likely effects of simulated forecast uncertainties in the trip end totals, since the constrained nature of the model ensures that they have a direct effect on the forecast trip distributions.

One way of examining these effects is to replace the O_i and D_j variables by random numbers from a normal distribution with a mean equal to the observed 1971 value and a standard deviation choosen to reflect various levels of forecast uncertainty. Table 4 gives the means and standard deviations of forecast year performances for both models using 1966 NLLS parameter estimates, based on 30 sets of O_i and D_j values generated for various levels of forecast uncertainty. For comparison purposes the results of simulating sampling errors in both O_i and D_j are also shown.

The results show that effects of forecast uncertainties in the trip end totals are considerably greater than the effects of assuming that

Table 4. Goodness-of-fit forecasts of 1971 with simulated errors in the trip end variables

Level of forecast error	Singly constrained: Mean $\hat{\sigma}$	Singly constrained: Standard deviation of $\hat{\sigma}$	Doubly constrainted: Mean $\hat{\sigma}$	Doubly constrainted: Standard deviation of $\hat{\sigma}$
None	9.3	–	5.4	–
Sampling errors in O_i, D_j	9.4	.21	5.0	.06
10 per cent of 1971 O_i, D_j	9.6	1.00	5.6	.49
20 per cent of 1971 O_i, D_j	10.7	1.67	7.3	1.44
30 per cent of 1971 O_i, D_j	12.3	2.51	9.5	2.63
40 per cent of 1971 O_i, D_j	14.0	3.55	12.1	4.00
50 per cent of 1971 O_i, D_j	14.4	3.10	13.1	3.31
60 per cent of 1971 O_i, D_j	15.8	3.86	15.4	4.29

1966 parameters can be used to predict the 1971 pattern of trips. Furthermore, it is noted that these forecasting uncertainties are certainly underestimated since it is very unlikely that the confidence limits for the forecasts will be normally distributed around the correct value. They will be biased in some way.

8.4.4 *On the significance of model errors and 'fiddle' or 'fudge' factors*

So far nothing has been said about the role of model errors as a factor affecting forecast performance. In the previous sections this error term has been ignored when making forecasts for 1971; however, it can be used to estimate crude confidence intervals for the model. For example, for the singly constrained model, forecasts are made for 1971 with the 1966 error terms being simulated by generating random normally distributed numbers with the appropriate 1966 standard deviation. Rerunning the procedure 30 times produced a mean $\hat{\sigma}$ for the model's fit to the 1971 data of 10.2, with a standard deviation of 0.02, compared with a $\hat{\sigma}$ of 9.3 when the error term is ignored.

It is also possible to examine the effects on forecast accuracy of treating the error term as a fiddle factor so that the model has a perfect fit to the 1966 data. Thus an additive fiddle factor, f_{ij}, is defined as

$$f_{ij}{}^{1966} = N_{ij}{}^{1966} - T_{ij}(\varphi_1)^{1966} \tag{4.1}$$

so that the modified 1971 forecast becomes

$$T_{ij}^{*} = T_{ij}(\varphi_1)^{1971} + f_{ij}{}^{1966} \tag{4.2}$$

This produces a residual standard deviation of 8.7 for the singly constrained model, a slight improvement. However, a more common fiddle factor is a multiplicative one:

$$f_{ij}{}^{1966} = N_{ij}{}^{1966} \,/\, T_{ij}(\varphi_1)^{1966} \tag{4.3}$$

so that the modified 1971 forecast becomes

$$T_{ij}^{*} = T_{ij}(\varphi_1)^{1971}\; f_{ij}{}^{1966} \tag{4.4}$$

Unfortunately, this produces a forecast residual standard deviation of 53.6 for the singly constrained model and is a good example of why fiddle or fudge factor should not be used. It is far more satisfactory to retain the error term and use it to provide confidence limits for the model's forecasts.

8.5 AN EVALUATION OF SOME ALTERNATIVES DESIGNED TO IMPROVE MODEL PERFORMANCE

8.5.1 *Joint calibration of the 1966 and 1971 data*

One means of improving the temporal stability of the model's parameters is to estimate values for φ_1 which optimise a single estimator based on both 1966 and 1971 data. This may be expected to average out the effects of temporal change on the model's parameters. For example, the joint 1966 and 1971 NLLS estimator involves minimising $F(\varphi_1)$:

$$F(\varphi_1) = \sum_i \sum_j (T_{ij}(\varphi_1)^{1966} - N_{ij}{}^{1966})^2 + \sum_i \sum_j (T_{ij}(\varphi_1)^{1971} - N_{ij}{}^{1971})^2 \qquad (5.1)$$

This approach is evaluated using the singly constrained model and the results obtained for the seven estimators are reported in table 5. It is not surprising that model performances are very similar to the results described in tables 2, 3 and 4, although the best performance

Table 5. Parameter valeus and goodness-of-fit for joint 1966 and 1971 calibrations

	Singly constrained model		
		1966	1971
Estimator	φ_1	$\hat{\sigma}$	$\hat{\sigma}$
MLE	.135	3.5	9.4
NLLS	.111	3.6	9.2
MSAE	.171	3.9	10.0
MMX	.074	4.4	9.8
MCS	.079	4.3	9.5
WLS1	.050	5.2	12.0
WLS2	.166	3.8	9.9

is now provided by the joint MLE estimator. Clearly, this approach would be far more useful if a third data set was available to test forecast performances.

8.5.2 *Some effects of alternative global deterrence functions*

The results obtained in sections 8.3 and 8.4 are based on models which use a negative exponential deterrence function with only one undetermined parameter. This function was choosen partly for convenience and partly because it is widely used in practice. However, the question now arises as to whether or not other global deterrence functions, some with more than one undertermined parameter, will produce a different interpretation. To test this hypothesis the following deterrence functions are investigated for the singly constrained model:

a negative power function

$$\beta(\varphi) = c_{ij}^{-\varphi_1}, \tag{5.2}$$

a simplified Tanner function

$$\beta(\varphi) = \exp(-\varphi_1 c_{ij}) c_{ij}^{-\varphi_2}, \tag{5.3}$$

a negative exponential function with a variable limit which the function approaches asymptotically as $c_{ij} \to \infty$

$$\beta(\varphi) = \varphi_1 + \exp(-\varphi_2 c_{ij}), \tag{5.4}$$

a modified Weibull function (Hyman and Gleave, 1976)

$$\beta(\varphi) = \exp(-\varphi_1 c_{ij}^{\varphi_2}), \tag{5.5}$$

the generalised function of March (1971)

$$\beta(\varphi) = \frac{\varphi_2 \varphi_3^{\varphi_1/\varphi_2} c_{ij}^{\varphi_1 - 1} \exp(-\varphi_3 c_{ij}^{\varphi_2})}{\Gamma(\varphi_1/\varphi_2)} \tag{5.6}$$

where $\Gamma(\ )$ is a gamma function,

an incomplete gamma function

$$\beta(\varphi) = \int_{c_{ij}}^{\infty} x^{\varphi_1 - 1} \exp(-x)dx, \tag{5.7}$$

and a form of the incomplete beta function

$$\beta(\varphi) = \int_{(c_{ij}^* - \varphi_3)^2}^{1} x^{2\varphi_1} (1-x)^{\varphi_2}\, dx, \tag{5.8}$$

where

$$c_{ij}^* = \frac{c_{ij}}{\sup\limits_{(i,j)\in G} c_{ij}}.$$

Table 6 shows the parameter values and forecast performance of these functions based on NLLS estimation, using a modified version

Table 6. Parameter values, goodness-of-fit, and forecast performances of some alternative deterrence functions

Deterrence function	Parameter values φ_1	φ_2	φ_3	Calibration $\hat{\sigma}$	Forecast* $\hat{\sigma}$
1966 data					
eq. 5.2	1.320			4.4	10.8
eq. 5.3	.135	– .049		3.5	9.3**
eq. 5.4	– .007	.128		3.5	9.4
eq. 5.5	.116	1.088		3.5	9.3
eq. 5.6	1.162	.957	.169	3.5	9.3
eq. 5.7	16.634			4.2	10.4
eq. 5.8	– .474	16.740	.001	3.5	9.3**
1971 data					
eq. 5.2	1.051			10.3	4.7
eq. 5.3	.045	.648		9.6	4.0
eq. 5.4	.003	.110		9.1	3.7
eq. 5.5	.115	.929		9.2	3.7
eq. 5.6	1.134	.828	.216	9.1	3.7
eq. 5.7	19.806			9.7	4.4
eq. 5.8	– .820	7.769	.147	10.3	19.6

Notes: * With 1966 data the forecast is of 1971, with 1971 data it is a back-forecast of 1966.
** Forecast performance is better than calibration performance due to local convexities in the sum of squares function.

of Harwell subroutine VAO7AD (Fletcher, 1971) to minimise the sum of squares functions with two or more undetermined parameters.

The numerical differences between the 1966 and 1971 parameters are greater than those recorded for the negative exponential functions. However, the levels of forecast performance are in nearly all cases very similar, confirming the interpretation that calibration and forecast year performances are about the same despite considerable numerical differences in the parameters. The results also show that the choice of alternative deterrence functions has relatively little beneficial effect on model goodness-of-fit.

8.5.3 *Towards a maximum performance spatial interaction model*

The question of whether or not model performance is adequate has yet to be resolved. Considering the philosophy of the problem, goodness-of-fit is important and it may be necessary to actively seek ways of improving the fit of the model. In so doing, it is important to avoid merely fitting outliers in the data, and to seek a parsimonious model since the fit of any model can be improved simply by adding more and more parameters to it. One approach is to experiment with deterrence function approximations which make no assumptions about the form β takes, except that it obeys certain rules throughout its domain of definition. However, it is reported elsewhere that even when the data are allowed to identify the most appropriate global deterrence function, by various series or orthonormal approximations, there is only a small improvement in model goodness-of-fit (Openshaw and Connolly, 1977).

It is now considered unlikely that model performance can be much improved by the identification of a more appropriate global deterrence function. The only remaining course of action is to relax the assumption that there is a single global function which can be defined for all trips. Thus the basic singly constrained spatial interaction model given in equation (2.1) can be re-written as

$$T_{ij}(\mathbf{\Phi}) = O_i D_j B_j(\mathbf{\Phi}) \beta_k(\boldsymbol{\varphi}_k : c_{ij}) \tag{5.9}$$

where

$$B_j(\mathbf{\Phi}) - \left[\sum_{k=1}^{m} \sum_{i \in H_k} O_i \beta_k(\boldsymbol{\varphi}_k : c_{ij}) \right]^{-1} \tag{5.10}$$

for all $(i, j) \in G_k$, $k = 1, 2, \ldots, m$

m	is the number of deterrence functions
G_k	is the set of pairs of origin- and destination-zone indices associated with the kth deterrence function
H_k	is the set of origin zones associated with G_k, i.e. $H_k \equiv \{i : (i, j) \in G_k\}$, and
$\beta_k(\boldsymbol{\varphi}_k : c_{ij})$	is the kth deterrence function characterised by the sequence of parameters $\{\varphi_{ik}\}\, i = 1, 2, \ldots, m_k$, where m_k is the number of parameters associated with the kth function and $\boldsymbol{\Phi}$ is the set of all these parameters.

In this model the β_k characterise a family of deterrence functions embedded within the framework of the same model, because of the trip end constraints. When there is only one function in the family, then equation (5.9) is the same as equation (2.1). However, this approach permits a high level of explicit spatial representation, since the trip-deterrence-function classification, G_k, $k = 1, 2, \ldots m$ can be structured in whatever manner is considered appropriate. Of course the problem now arises as to how the members of G are to be defined.

One possible trip-deterrence-function classification is to allocate the trips going to each destination zone a different deterrence function. For example, Wilson (1973) allocates each origin or destination zone a different negative exponential deterrence function. Thus with 73 destination zones there will be 73 undetermined parameters in Φ. However, in many studies there is no basis for an a priori specification of which trips belong to which deterrence function, although if the zoning system is fine enough this may be possible, for example, based on various notions of central place theory or urban system structure. It is necessary, therefore, to use a heuristic procedure to allocate trips to one of a family of functions, such that the overall performance of the model tends towards some upper limit. In this way it is possible to identify a maximum performance model for a given family of deterrence functions (Openshaw and Connolly, 1977). Table 7 shows the results obtained with a singly constrained model using families of negative exponential functions.

Table 7. Goodness-of-fit of some families of deterrence functions

Number of functions in family	Calibrations 1966	Calibrations 1971	Forecasts of 1966	Forecasts of 1971
1 (Conventional model)	3.5	9.1	3.5	9.3
2	1.7	4.9	2.7	8.5
3	1.3	4.1	2.5	8.7
4	1.2	2.6	2.5	8.8
5	1.1	2.8	2.4	8.4
6	.7	2.2	2.4	16.9
73 (After Wilson, 1973)	2.7	6.0	3.4	8.4

Compared with the conventional global deterrence function model, the use of a family of six negative exponential functions provides a tremendous improvement in model goodness-of-fit in both 1966 and 1971. However, there is only a small gain in forecast performance. Indeed it is interesting to observe that for the first time the model's forecasts appear to be sensitive to the use of 1966 parameters. Moreover, forecast performance is now far worse than calibration year performance and this can be largely attributed to the static parameters. For example, the six family model calibrated in 1966 with a simultaneous estimation of parameters and the spatial domain of each function has a forecast residual standard deviation of 16.9 for 1971. Part of this poor forecast performance can be attributed to the use of a 1966 trip-deterrence-function classification, but mainly it is a result of inappropriate parameters. If the model is re-calibrated but retains the 1966 definition of which trips belong to which functions then the forecast residual standard deviation reduces to 7.4.

8.5.4 *Approximately optimal zoning systems*

A final factor to consider is the problem of zone design. In model building it is conventional practice to assume that the data being modelled is fixed apart from sampling errors, and attention is mainly focused on the statistical problems associated with the estimation of the model's undetermined parameters. However, when the data being studied relate to spatial zones and have been spatially aggregated one or more times, the data are no longer unique but are dependent on the choice of a particular zoning system. So in addition to sampling error there is a modifiable unit problem. Thus for thus study the 73 zones covering the study region are formed by aggregating interaction data from a very fine zoning system based on census enumeration districts of which there are several thousands. Obviously there are a tremendously large number of alternative arrangements of this data into 73 zones. Moreover, the selection of one particular aggregation of 73 zones and the decision that there shall be 73 zones are both arbitrary. This also raises the fundamental statistical problem concerning precisely what is the population about which inferences are being made from one aggregation of sample data?

It is now known that the goodness-of-fit and the parameter values of spatial interaction and other models are dependent on the choice of zoning system, and thus implicitly on the scale and level of aggregation of the data (see Openshaw, 1977a). Thus it can be argued that whenever spatially aggregated data are being studied there is an interdependence between the zoning system and the model and its parameters. It is a two way relationship. One approach to this problem is to attempt to simultaneously estimate the most appropriate parameter values for the model as well as the most appropriate zoning system for the model by using a common estimator (see Openshaw, 1977b). Thus the purely spatial equivalent of parameter estimation is to improve model goodness-of-fit by seeking a zoning system to fit a given model, thereby avoiding possibly poor levels of fit resulting from the unfortunate choice of a 'poor' zoning system. This is analagous to a statistician carefully selecting an estimator for a particular model assuming fixed spatial data.

For present purposes it is of interest to investigate the extent to which the choice of approximately optimal zoning systems offer any improvement to the spatial interaction model's forecast performance. In this exercise the 73 original zones are re-aggregated to form 67, 61, 51, 42, 32, and 22 zone arrangements. Each of these zoning systems is designed to optimise the fit of a singly constrained model with a single negative exponential deterrence function. In this instance the value of φ_1 is held fixed at the 73 zone, NLLS, 1966 value and the fit of the model improved only by changing the zoning system, using an algorithm described in Openshaw (1977c). The zoning system and model parameter are then held constant and used to forecast the 1971 trip pattern (see Table 8).

Although the optimal fit zoning systems provide a considerable improvement in model performance in the calibration year, their

Table 8. Performance of some approximately optimal fit zoning systems

Number of zones	Forecasts of 1971 using 1966 optimal zoning systems, $\hat{\sigma}$	Forecasts of 1971 using a random zoning system, $\hat{\sigma}$	Optimal fit 1971 zoning system, $\hat{\sigma}$
22	82.2	83.2	34.6
32	35.0	40.4	24.4
42	24.2	24.5	17.3
51	18.2	17.8	12.2
61	11.3	11.2	10.9
67	11.2	10.4	9.6

forecasts of the 1971 situation are no better than those based on purely randomly generated zoning systems with an equivalent number of zones. This may indicate that the optimal zoning systems are not stable over time. However, a large part of the blame for poor performance can be attributed to the use of inappropriate parameters. For example, when the model is re-calibrated for the best-fit 1966, 22 zone system but using 1971 data, the residual standard deviation is reduced from 82.2 to 41.7.

8.6 CONCLUSIONS

This paper has examined the effects of a number of alternative ways of estimating what is basically the same relationship in terms of calibration and forecast performance. The results show that the forecast performance of the conventional model is virtually the same as that obtained during calibration, provided that accurate estimates of the O_i, D_j, and c_{ij} variables are available. This suggests that the model can be validated in terms of its calibration year performance and that only a small error is introduced into the conventional model by assuming static parameters in a forecasting context, compared with the errors in the exogenous variables O_i, D_j, and c_{ij}.

This does not mean that the model's performance is satisfactory. Indeed the residual standard deviations are very large in relation to the size distribution of the observed trips. Furthermore, it is apparent that the model's performance is largely insensitive to a wide range of parameter values in a number of different deterrence functions, because it appears to be dominated by errors related to the specification of the model itself. This error component is large and essentially random, hence the model's apparent insensitivity to its static parameters in a forecasting context and to a wide range of different parameter values and different deterrence functions in a calibration situation. The origin and destination constraints obviously contribute, but the fact that model errors are to blame can be clearly seen in the increased sensitivity of the model to its parameters when attempts are made to improve model performance.

There is a trade-off to be made between a conventional model, which is insensitive to parameter changes and has relatively poor calibration and a similar forecast performance, and alternative forms based on either families of deterrence functions or optimal zoning systems, which have a far better calibration year performance but poor forecast accuracy due to an increased sensitivity to parameter changes. If one accepts that a high level of forecast uncertainty is inherent in all exogenous forecasts of the O_i and D_j variables, then clearly current spatial interaction models should only be used for calibration year policy testing. No matter how much the model itself can be improved it is unlikely that more accurate O_i and D_j forecasts will be available for a long time. However, in the calibration year there is no forecast uncertainty, the assumption of static parameters is avoided, and model performance can be improved by the use of different deterrence function characterisations or by the use of approximately optimal zoning systems or by both.

It is concluded, therefore, that the results of this study indicate that policy testing in the calibration year is the most appropriate context for the applied use of spatial interaction models, and that any 'real world' interpretation of the model output should preferably be avoided or heavily qualified (see Openshaw, 1977d, for a discussion). It is also noted that there is a desperate need for further research into the statistical and empirical aspects of spatial interaction models, especially into those problems which result from the special nature of spatial interaction data.

REFERENCES

Bard, Y. (1974), *Nonlinear parameter estimation*, Academic Press, New York.

Batty, M. and Mackie, S. (1972), 'The calibration of gravity, entropy, and related models of spatial interaction', *Environment and Planning*, 4, 205–233.

Batty, M. (1976), *Urban modelling: algorithms, calibrations, predictions*, Cambridge University Press, Cambridge.

Black, J. A. and Salter, R. J. (1975), 'A statistical evaluation of the accuracy of a family of gravity models', *Proceedings of the Institute of Civil Engineers*, Part 2, 59, 1–20.

Byler, J. and O'Sullivan, P. (1974), 'The forecasting ability and temporal stability of the coefficients of gravity models applied to truck traffic', *Traffic Engineering and Control*, 15, 474–476.

Evans, A. W. (1971), 'The calibration of trip distribution models with exponential or similar cost functions', *Transportation Research*, 5, 15–38.

Fletcher, R. (1971), *A modified Marquardt subroutine for non-linear least squares*, AERE R 6799, Her Majesty's Stationary Office, London.

Gough, A. J. (1976), 'Social physics and local authority planning', in *Housing and Class in Britain*, Russel Press, London, 87–104.

Hathaway, P. J. (1975), 'Trip distribution and disaggregation', *Environment and Planning A*, 7, 71–97.

Hopper, M. J. (1973), *Harwell subroutine library: a catalogue of subroutines*, AERE R 7477, Her Majesty's Stationary Office, London.

Hyman, G. M. and Gleave, D. (1976), 'A reasonable theory of migration', *Centre for Environmental Studies*, RP-22, London.

Leese, M. N. (1977), 'A statistical test for trip distribution models', Paper presented at Transport Study Group Conference, University College London.

Madsen, K. (1973), 'An algorithm for a minimax solution of over-determined systems of non-linear equations', *Technical Paper* 559, Harwell, Berks.

March, L. (1971), 'Urban systems: a generalised distribution function', in *Urban and Regional Planning*, A. G. Wilson (ed.), Pion, London.

Openshaw. S. (1975), *Some theoretical and applied aspects of spatial interaction shopping models*, CATMOG, 4, Geo. Abstracts, Norwich.

Openshaw, S. (1976), 'An empirical study of some spatial interaction models', *Environment and Planning A*, 8, 23–41.

Openshaw, S. (1977a), 'A geographical solution to scale and aggregation problems in region building, partitioning, and spatial modelling', *Transactions of the Institute of British Geographers, New Series*, 2, 459–472.

Openshaw, S. (1977b), 'Optimal zoning systems for spatial interaction models', *Environment and Planning A*, 9, 169–184.

Openshaw, S. (1977c), 'An optimal zoning approach to the study of spatially aggregated data', in I. Masser and P. J. B. Brown (eds.), *Spatial representation and spatial interaction*, Martinus Nijhoff, Leiden, 95-113.

Openshaw, S. (1977d), 'Models in planning: a state of the art review', *Planning Outlook*, 20, 30–34.

Openshaw, S. and Connolly, C. J. (1977), 'Empirically derived deterrence functions for maximum performance spatial interaction models', *Environment and Planning A*, 9, 1067–1079.

Rice, J. R. and White, J. S. (1964), 'Norms for smoothing and estimation', *SIAM Review*, 6, 243–255.

Sayer, R. A. (1976), *A critique of urban modelling*, Progress in Planning, 6, Part 3, Pergamon, Oxford.

Schlossmacher, E. J. M. (1973), 'An iterative technique for absolute deviation curve fitting', *Journal of the American Statistical Association*, 68, 857–859.

Snickars, F. and Weibull, J. W. (1977), 'A minimum information principle: theory and practice', *Regional Science and Urban Economics*, 7, 137–168.

Stetzer, F. (1976), 'Parameter estimation for the constrained gravity model: a comparison of six methods, *Environment and Planning A*, 8, 673–683.

Wilson, A. G. (1967), 'A statistical theory of spatial distribution models', *Transportation Research*, 1, 253–269.

Wilson, A. G. (1970), *Entropy in urban and regional modelling*, Pion, London.

Wilson, A. G. (1973), 'Further developments of entropy maximising transport models', *Transportation Planning and Technology*, I, 183-193.

9. Two estimation methods for singly constrained spatial distribution models

JAN VAN EST and JAN VAN SETTEN*

9.1 INTRODUCTION

Research in physical planning often uses distribution models of the form of gravity-, entropy- or logit-models for the description of spatial phenomena like residential and shopping interactions (Wilson, 1974). These distribution models differ according to the constraints being imposed. Generally, a distinction can be made between models with one constraint and those with two. This paper concentrates on the distribution model with one constraint, the so-called 'singly-constrained' distribution model.

The distribution model here to be considered describes interactions which will be influenced by several factors, e.g. attraction and distance variables. The influence of explanatory variables finds expression in the accessory parameters. The estimation of the values of these parameters and the testing of their significance is called 'calibration'. The estimation and testing of the parameters are two statistically strongly related procedures. Several methods of estimation are possible. The two best-known procedures, i.e. the 'maximum likelihood' and the 'least-squares' method are considered in this paper.

The application of the maximum likelihood method requires an iterative solution procedure, because of the intrinsic non-linearity of the likelihood function. This procedure requires explicit consideration of the conditions for the existence of a unique solution. Depending on the introduction of disturbances as a specification error and the origin constraints, the least-squares method may also imply a non-linear estimation procedure (Openshaw, this volume). However, it will be shown that a procedure based on linear least

*Planologisch Studiecentrum TNO, P.O. Box 45, Delft, The Netherlands.

squares can also be appropriate. In this paper the maximum likelihood estimator for the parameters of a singly-constrained distribution model (section 9.3) will be compared with the linear estimator (section 9.4). An application to a particular data-set is presented in section 9.5.

9.2 THE CALIBRATION OF A MODEL

The calibration of a model consists of using:
1. Methods to estimate the values of unknown parameters of the model;
2. Methods to test whether:
   a. the individually estimated parameters of the model differ significantly from zero,
   b. the calibrated model is able to represent the original situation, which generated the data, in a satisfying way.

In distribution models, especially in English studies, (2.a) has received little or no attention (an exception is Openshaw, this volume). For example, Wilson (1974) and Batty (1976) concentrate very much on estimation techniques and the testing of the total model.

In this paper tests for separate parameter estimates will also receive attention. The method of least squares gives, if the well-known assumptions are satisfied, an estimator which will be normally distributed. Therefore, the estimated standard deviations and/or the *F*-test can be used to investigate whether each independent variable and the set of all regressors have a significant influence on the phenomenon to be described. At the same time, the coefficient of determination can be used to test the linear relationship of the variables concerned (see inter alia Johnson, 1972). The maximum likelihood method gives a similar result because the estimator is asymptotically normally distributed (see Theil, 1971). Tests for the significance of (separate) parameters nevertheless have the objection that in the case of large numbers of interactions (the degree of freedom) they may not be sufficiently discriminating. An increase in the significance of one variable from 99.6 to 99.8 due to extra observations is not easily interpretable. In other words, for spatial distribution matrices involving large numbers of interactions the *F*-tests are relatively insensitive, especially when there are several parameters with a large variation in *F*-values.

Therefore, overall tests are also desired; they will help one to pass judgement concerning the model. For the distribution model it is possible to think of a comparison between the observed and the estimated values of an interaction matrix. With regard to the goodness-of-fit of the complete model, some possible indicators will be mentioned.

The simplest indicator is the coefficient of determination, which may be defined for the observed and estimated values of the interaction matrix (R^2 flows in table 1). This statistic is then a measure for the explained variance of the model. Comparison of the observed (N) and estimated (R) interaction matrix, can also be effected by using the method of the least squares:

$$R_{ij} = a + bN_{ij} \qquad i, j = 1, \ldots, n \tag{2.1}$$

where N_{ij} and R_{ij} are the typical elements respectively. When the conformaty of the calculated and observed values is complete, then $a = 0$ and $b = 1$. By application of the least-squares method, coefficients a and b can be estimated and in the same way tested for significant deviations from zero and one respectively. The linear relationship of the variables in equation (2.1) can be investigated with

$$T_1 = 1 - \frac{\sum_i \sum_j (R_{ij} - R_{ij}^*)^2}{\sum_i \sum_j (R_{ij} - \bar{N})^2} \tag{2.2}$$

in which R_{ij}^* is the estimated R_{ij} in equation (2.1) and $\bar{N}$ is the average of the observed N_{ij}. This statistic can be used as a test for the linear relationship between the observed and estimated values. The method of least squares for comparing the estimated and observed values of the interaction matrices is not only a simple method, it also has, in general, useful properties. However, the number of zones, as usually is the case with spatial distribution models, may be too large. Then one can take a-select partial samples. This procedure has been described by Stopher (1975).

A commonly used measure of goodness-of-fit in spatial interaction models is the difference between the observed and estimated mean trip length (Batty, 1970). It can be represented as

$$T_2 = \sum_i \sum_j (R_{ij} d_{ij} - N_{ij} d_{ij}) \tag{2.3}$$

in which d_{ij} represents the distance measure for the spatial separation of zones i and j. Independent of the distance distribution of the interactions, calibration of the model leads in general to values of T_2 which are equal or quite near to zero. For this reason T_2 is often used as a termination criterion for the iterative procedure in the maximum likelihood method.

A remark concerning the mean trip length in a distribution model is in order. Whether the mean trip length of the interactions can be used as a test statistic depends on the deterrence function of the gravity model. For specific choices of the deterrence function the maximum likelihood method automatically produces a value of T_2 equal to zero. The mean trip length is then useless as a test statistic. Consider a general formulation of the deterrence function

$$f(d_{ij}) = \exp \{ \beta h(d_{ij}) \} \tag{2.4}$$

Setting the partial derivative of the likelihood (see section 9.3, equation [3.6]) with respect to β equal to zero yields

$$\sum_i \sum_j N_{ij} h(d_{ij}) = \sum_i \sum_j R_{ij} h(d_{ij}) \tag{2.5}$$

From equation (2.5) it then follows that for a linear form of the transformation function h, T_2 will equal zero. Only when h is non-linear can T_2 be used as a test statistic. Therefore, it seems advisable to use in all cases the whole distribution instead of the mean for testing procedures.

For the present purpose only two measures of goodness-of-fit have been mentioned. Alternative indicators can be found in van Est and van Setten (1977). In the following sections it will be investigated how to use maximum likelihood and least-squares estimation techniques in order to obtain parameter estimates and significance tests for the singly constrained multiplicative interaction model.

9.3 THE MAXIMUM LIKELIHOOD METHOD

The likelihood is defined as the joint probability function of the observed random variables, which is considered to be dependent on the parameters. When this function is sufficiently often differentiable, it is possible to maximize it with respect to the parameters in a simple way. The procedure is to apply a logarithmic transformation (invariant w.r.t. maximization), and then to put the partial derivatives to zero and solve the equation system for the unknown parameters. Finally, the second-order conditions for a maximum ought to be satisfied. The 'solutions' obtained in the foregoing way are the maximum likelihood (ML) estimates. These estimators have important asymptotic properties (see Theil, 1971).

Starting from the single constraint over the origins, the multiplicative distribution model for the interactions reads as

$$R_{ij} = R_{i.} \frac{\exp\{\alpha' X_j + \boldsymbol{\beta}' Y_{ij}\}}{\sum_k \exp\{\alpha' X_k + \boldsymbol{\beta}' Y_{ik}\}} \quad i, j = 1, \ldots, n \qquad (3.1)$$

in which

R_{ij} = the number of interactions between zone i and j

X_j = a vector of attraction variables for zones j

Y_{ij} = a vector of distance variables for the spatial separation between zones i and j (deterrence function)

R_i = the total number of interactions, generating at zone i

n = the number of zones

α = a vector of unknown attraction parameters

$\boldsymbol{\beta}$ = a vector of (unknown) spatial separation parameters.

Note that only the ratio's $R_{ij} / R_{i.}$ are modelled. It is assumed that $R_{i.}$ is determined exogenously and is equal to the observed total number of interactions from zone i. In the calibration methods to be discussed there will be in fact no formal difference between the attraction vector, X_j, and the spatial separation matrix, Y_{ij}. Therefore a generalized notation may be more convenient:

$$R_{ij} = R_{i.} \frac{\exp\{\boldsymbol{\gamma}' V_{ij}\}}{\sum_k \exp\{\boldsymbol{\gamma}' V_{ik}\}} \qquad (3.2)$$

in which

$$\boldsymbol{\gamma}' = (\alpha', \boldsymbol{\beta}')$$

$$V_{ij} = (X_j', Y_{ij}')'$$

The distribution model of equation (3.1) may also be interpreted as a probability model. The conditional probability of an interaction between zones i and j is defined as

$$p_{ij} = \frac{R_{ij}}{R_{i.}} \tag{3.3}$$

The unconditional probability of an interaction between zones i and j is then defined as

$$q_{ij} = \frac{R_{i.}}{R} \cdot p_{ij} \tag{3.4}$$

where R = the total number of interactions. See for instance Batty (1976) and Williams (1974). Note that p_{ij} and q_{ij} differ only by a multiplicative constant, since $R_{i.}$ and R are determined exogenously. Denoting the observed interaction matrix with typical element N_{ij}, as N the likelihood function to be derived is

$$L(\boldsymbol{\gamma}) = \frac{N!}{\prod_i \prod_j N_{ij}!} \prod_i \prod_j q_{ij}(\boldsymbol{\gamma})^{N_{ij}} \tag{3.5}$$

since the joint probability distribution of the number of interactions is multinomial.

The method of the maximum likelihood tries to find estimates for the probabilities p_{ij} i.e. $\boldsymbol{\gamma}$'s by maximizing the logarithm of the likelihood function:

$$\underset{\gamma}{\text{MAX}}\ ln\, L(\boldsymbol{\gamma}) = \underset{\gamma}{\text{MAX}} \{C + \sum_i \sum_j N_{ij}\, ln\, p_{ij}\} \tag{3.6}$$

in which the constant term, C, is independent of $\boldsymbol{\gamma}$ and the fact has been used that $ln p_{ij}$ and $ln q_{ij}$ only differ by an additive constant.

In order to keep the notation simple, it is not indicated explicitly that p_{ij} is a function of $\boldsymbol{\gamma}$. To reach a maximum of lnL the first-order derivatives must be equal to zero and the Hessian, i.e. the matrix of the second-order derivatives, must be negative definite in the stationary point. For the model in question this necessary condition is normally satisfied. Moreover, it appears that the function to be maximized is concave, so that the derived maximum is also the absolute maximum. The derivatives for the qth component of γ are

$$\frac{\partial \, lnL}{\partial \gamma^q} = \sum_i \sum_j N_{ij} V^q_{ij} - \sum_i (N_{i.}) \sum_j p_{ij} V^q_{ij} \tag{3.7}$$

where V^q_{ij} = the qth component of V_{iq}. The second-order derivatives then build a matrix, in which the typical element is

$$\frac{\partial^2 \, lnL}{\partial \gamma^p \, \partial \gamma^q} = - \sum_i N_{i.} \, \{ \sum_j p_{ij} \left[V^p_{ij} - \sum_k p_{ik} V^p_{ik} \right] V^q_{ij} \} \tag{3.8}$$

or

$$\frac{\partial^2 \, lnL}{\partial \gamma^p \, \partial \gamma^q} = - \sum_i N_{i.} \, \{ \sum_j p_{ij} \left[V^p_{ij} - \sum_k p_{ik} V^p_{ik} \right] \left[V^q_{ij} - \sum_k p_{ik} V^q_{ij} \right] \tag{3.9}$$

because

$$\sum_j p_{ij} \left[V^p_{ij} - \sum_k p_{ik} V^p_{ik} \right] \left[\sum_k p_{ik} V^q_{ik} \right] = 0 \tag{3.10}$$

when

$$a^p_{ij} \equiv V^p_{ij} - \sum_k p_{ik} V^p_{ik} \tag{3.11}$$

then in vector notation, the Hessian will be

$$\boldsymbol{H} = - \sum_i N_i \sum_j p_{ij} \boldsymbol{a}_{ij} \boldsymbol{a}'_{ij} \tag{3.12}$$

The quadratic form, $\boldsymbol{x}'\boldsymbol{H}\boldsymbol{x}$, of this Hessian will be negative of zero for every $\boldsymbol{\gamma}$. The latter case can only appear in two ways (provided $\boldsymbol{x} \neq 0$):

1. if there is multicollinearity within all matrices of $\boldsymbol{a}_{ij}\boldsymbol{a}_{ij}'$, $j = 1, \ldots, n$;
2. if vectors V_{ij}^{p} equal their calculated mean for each j, $j = 1, \ldots, n$.

Both possibilities are practically excluded so that the Hessian is negative definite indeed. The Hessian is not only negative definite in the optimum, but this property holds for the whole parameter range. Hereby the function is concave. Because the function must have at least one point of stationarity the function has an absolute maximum. This property is of importance for optimization methods. The intrinsic non-linearity of the likelihood function makes an interative procedure necessary. Because of the previous mentioned function properties, converging optimalization methods lead to estimations defined by the estimation method, a solution which is statistically convergent to the true values of the parameter vector.

9.4 THE LEAST-SQUARES METHOD FOR THE SINGLY-CONSTRAINED MODEL

The model of equation (3.1) is not yet complete; a disturbance term for all variables which are not taken into consideration should be added. Because the distribution model is multiplicative, this specification error is also added multiplicatively, in such a way that the additivity constraints are satisfied. In the generalized notation of equation (3.2) the model reads as

$$N_{ij} = N_{i.} \frac{\exp\{\boldsymbol{\gamma}'\boldsymbol{V}_{ij} + \mu_{ij}\}}{\sum_k \exp\{\boldsymbol{\gamma}'\boldsymbol{V}_{ik} + \mu_{ik}\}} \tag{4.1}$$

For a particular estimator, $\hat{\boldsymbol{\gamma}}$, of the parameter, then holds:

$$N_{ij} = N_{i.} \frac{\exp\{\hat{\boldsymbol{\gamma}}'\boldsymbol{V}_{ij} + \epsilon_{ij}\}}{\sum_k \exp\{\hat{\boldsymbol{\gamma}}'\boldsymbol{V}_{ik} + \epsilon_{ik}\}} \tag{4.2}$$

where ϵ_{ij} is a residual term, determined implicitly by (4.2). The disturbance term in model (4.1) is assumed to be normally distributed for all vectors $\boldsymbol{\mu}_i = (\mu_{i1}, \ldots, \mu_{in})$, independently and identically (homoscedasticity) with zero expectation and bounded variance. The denominator of equation (4.2) is only a function of i, so that (4.1) can be written as

$$N_{ij} = N_{i.}\delta_i \exp \{\hat{\boldsymbol{\gamma}}' V_{ij} + \epsilon_{ij}\} \tag{4.3}$$

A logarithmic transformation of equation (4.3) then gives

$$ln\, N_{ij} = ln\, N_{i.}\delta_i + \hat{\boldsymbol{\gamma}}' V_{ij} + \epsilon_{ij} \tag{4.4}$$

Although equation (4.4) looks similar to a linear model, it is not linear in the parameters because $\boldsymbol{\delta}_i$ is still a function of these parameters:

$$\boldsymbol{\delta}_i \equiv \boldsymbol{\delta}_i\ (\alpha, \boldsymbol{\beta}). \tag{4.5}$$

However, linear least squares may still be applied if we proceed as follows.

When all variables in (4.4) for each origin are measured in deviation from their arithmetic mean, then (4.4) yields

$$ln\, N_{ij} - \frac{1}{n} \sum_m ln\, N_{im} = \hat{\boldsymbol{\gamma}} (V_{ij} - \frac{1}{n} \sum_m V_{im}) + \epsilon_{ij} - \frac{1}{n} \sum_m \epsilon_{im} \tag{4.6}$$

Since the distribution model of (4.2) is homogeneous of degree zero with respect to exp $\{\epsilon_{ij}\}$ for $i, j = 1, \ldots, n$, the level of ϵ_{ij} is not uniquely determined. Without any loss of generality, a zero mean for the residuals of each origin might be assumed, so that the average residual term in (4.6) vanishes.

Thus the standard linear model is derived, so that the ordinary least-squares procedures can be applied (see Johnston, 1972).

In most gravity modelling applications the denominator in equation (4.2) is normally eliminated by taking the ratio between N_{ij} and N_{im}, so that after a logarithmic transformation we obtain

$$ln N_{ij} - ln N_{im} = \hat{\gamma} (V_{ij} - V_{im}) + (\epsilon_{ij} - \epsilon_{im}) \tag{4.7}$$

In applying the least-squares method for the following has to be minimized:

$$\underset{\gamma}{\text{Min}} \sum_i \sum_j \sum_m (\epsilon_{ij} - \epsilon_{im})^2 . \tag{4.8}$$

Blokland (1975) has proved that minimization of (4.8) is equal to the minimization of

$$\text{Min} \sum_i \sum_j (\epsilon_{ij} - \overline{\epsilon}_i)^2 = \underset{\gamma}{\text{Min}} \sum_i \sum_j \epsilon_{ij}^2 \tag{4.9}$$

which follows if advantage is taken of the invariance of the level of $\overline{\epsilon}_i$. It can easily be seen that by applying (4.9) to (4.7) the same estimator as derived from (4.6) will be obtained.

Until now it has been assumed that the parameters are independent of the origin zones. As is common practice in regional science, this assumption would have to be very often relaxed to reflect zonal dependent parameters. Then model (4.6) can be presented as

$$ln N_{ij} - \frac{1}{n} \sum_m ln N_{im} = \hat{\gamma}_i (V_{ij} - \frac{1}{n} \sum_m V_{im}) + \epsilon_{ij} \tag{4.10}$$

Again the ordinary least-squares method can be applied to each row of the interaction matrix separately. For testing both models (4.6) and (4.10) readers are referred to van Est and van Setten (1978).

In equation (4.6) and (4.10) two different models have been presented. It has been shown that after suitable transformation of the variables N and V the ordinary least-squares method can be applied to both multiplicative distribution models. Apparently, the balancing factor δ_i of the distribution model may be interpreted as a constant term in a regression equation. The derivation of the least squares estimator has been performed through operations on the residual vector (see Johnston, 1972). Therefore it was not necessary to make an additional assumption about the disturbance term. Blokland (1975) used an alternative approach, but started with the disturbances. Then it is, just as in the linear distribution model (Barten, 1969), necessary to assume a zero mean of the disturbance term in order to satisfy the constant. However, this implies that the covariance matrix of the disturbances becomes singular. Blokland (1976) and Nakanishi and Cooper (1974) solved this singularity problems by eliminating one observation from every column. As has been shown, the approach which starts from properties of the residuals circumvents singularity problems automatically.

9.5 NUMERICAL RESULTS

9.5.1 *Introduction*

In the foregoing sections the methods of maximum likelihood and least-squares have been discussed. In this chapter a comparison will be made between both statistical methods. The comparison, however, will be restricted to an empirical point of view. For a singly constrained distribution model both of the methods will be applied. The data for the comparison refer to the agglomeration Eindhoven with a coarse zoning system of 12 zones. The application of the model has been based on the work done in the so-called CADSS-project (van Est, 1976). The attention is mainly focused on the following subjects:

- the result of the calibration of both models;
- the comparison of both statistical methods;
- conclusions.

9.5.2 *Results of the singly-constrained model*

The comparison between the method of maximum likelihood (ML) and the method of least squares (LS) is based upon a shopping model for the agglomeration Eindhoven. The VODAR-survey (Buro Goudappel, 1970) supplied the empirical data for the interaction matrix. The distance matrix resulted from a 'shortest route' algorithm over the traffic network and is measured in time units. The floor area of each shopping centre, measured in square meters, has been taken as the attraction factor for the shopping behaviour. The formal representation of the shopping model is

$$R_{ij} = R_{i.} \frac{A_j^{\alpha} e^{-\beta d_{ij}}}{\sum_k A_k^{\alpha} e^{-\beta d_{ik}}} \tag{5.1}$$

where

R_{ij} = the number of shopping trips between zone i and zone j

A_j = the floor area of the shopping centre in zone j

d_{ij} = the time distance between zone i and zone j

α, β = unknown parameters, to be estimated.

The results of the calibration are presented in table 1.

Table 1. Calibration results of the shopping model

	Least squares		Maximum Likelihood	
	Attraction	Distance	Attraction	Distance
Estimate	0.66	−0.0021	0.67	−0.0018
Standard error	0.08	0.0002	0.004	0.00002
F-value	60.2	168.5	25521	10582
R^2 (flows)	0.878		0.884	
T_1	0.857		0.855	
T_2	-21.5		0.001	

Although both methods, least squares and maximum likelihood, are fundamentally different, the figures in table 1 show that most results are similar. However, the differences between the F-values of both methods are quite large. This point of discussion will be dealt with in paragraph 9.5.3.

In this example of a shopping model, the F-value may be used as a test statistic, although with the maximum likelihood method this statistic seems not to be very fruitful. Moreover, T_2 cannot be used as a test statistic; it is only a measure of the attained accuracy of the approximated gradient in the optimization process (cf. Section 9.2).

9.5.3 *Comparison of both statistical methods*

With respect to the results of the calibration one might conclude that the estimates for each model are rather similar. Still both methods are fundamentally different. The method of least-squares is based on (the normal distribution of) the disturbances in equation (4.6) while the method of maximum likelihood has been based on a multinomial distribution of the allocations. The property of consistency, however, implies that both estimators converge in probability to the true parameters. So when the sample size is large enough both methods should show similar results. Nevertheless the differences between the standard errors of the parameters are obvious. It will be argued that for both methods it is important to take into account the number of origins respective of the number of interactions.

Applying the method of least-squares, the estimate of the variance of the disturbance term is normally based on the squared residuals about the least-squares line from equation (4.6):

$$s^2 = \frac{\epsilon'\epsilon}{n^2 - Q} \tag{5.2}$$

where $n^2 - Q$ equals the degrees of freedom. Because all the observations, i.e. the number of cells in the interaction matrix, will be taken into account, the degrees of freedom is equivalent with the number of observations (cells) minus the number of independent variables. This implies that in this case only the number of equations, i.e. the number of zones, defines the sample size, so that the number of interactions is unfortunately of no importance. In other words, the standard error is invariant for interaction samples; increasing the size of the interview sample does not necessarily improve the estimation, while the zoning system has an influence on the accuracy of the estimation.

The results for the method of least-squares are compared with the method of maximum likelihood. In this case the sample size does not depend on the number of zones, but is defined by the number of interactions (i.e. degrees of freedom). All elements of the Hessian in section 9.3 are a weighted sum of the interactions. From this it follows that a multiplication of all elements of the interaction matrix with a constant (say C) yields the same multiplication effect for the Hessian, so that the standard error will be decreased by a factor $\sqrt{C}$. Taking into account the sample definitions for both methods the resulting differences in the calibrated standard errors can now be interpreted very easily. As related to the shopping model, the standard error for the interaction model in the case of least-squares has to be better because of the squaring of the sample size. The standard errors in the case of maximum likelihood are in general so small, due to the sample size, that they do not have a much discriminating effect. However, it should be borne in mind that both methods start from different assumptions, e.g. the least-squares estimator takes a possible specification error into account.

A last remark deals with matrix cells with small observation values. Though from a statistical point of view they are unwished, zero cells sometimes are unavoidable. Providing there are no zero observed origin totals or zero attraction values in the ML case, the multinomial distribution of (3.6) can deal with zero N_{ij} and those cells do not prevent application of the method. On the contrary the partial logarithmic transformation of (4.6) does not allow for zero (N_{ij}) observations. Such a situation can also appear if there are zones missing observations (Bovy, 1976).

9.6 CONCLUSIONS

For an application of the linear least-squares method to the estimation of spatial distribution models there appears to be more opportunity than it is sometimes thought. For the singly-constrained trip distribution model the method is easy and can be used directly after a special logarithmic transformation without taking into account additivity constraints and without an iterative procedure. Because of that logarithmic transformation the balancing factors can be interpreted as a constant term (for each i). However, problems arise because only positive number are allowed now (a 'sound' model formulation and a good zoning system can overcome this problem). Moreover, the estimates for parameters and variance of the disturbance term are unbiased and F-tests can be performed very well.

The maximum likelihood method has, from a statistical point of view, very useful properties. Although an interative procedure is now needed, requirements for efficient optimization techniques are satisfied owing to the proved unicity of the solution. In contrast to the method of least-squares, zero observations are allowed. The experiences as mentioned in this paper did not lead to the rejection of the method of least-squares, nor of the method of maximum likelihood; each method has its advantages and limitations.

REFERENCES

Barten, A. P. (1969), 'Maximum likelihood estimation of a complete system of demand equations', *European Economic Review*, 9, 7-73.

Batty, M. (1970), 'Some problems of calibrating the Lowry model', *Environment and Planning*, 2, 95-144.

Batty, M. (1976), *Urban Modelling, Algorithms, Calibration, Predictions*, Cambridge University Press, Cambridge.

Blokland, J. (1975), 'Naar een model voor het afstandsgedrag', in F. Le Clercq, et al. (eds.), *Transportation Planning Research Colloquium*, Delft, 233-262.

Blokland, J. (1976), *Continuous Consumer Equivalence Scales*, Stenfert Kroese, Leiden.

Bovy, P. H. L. (1976), 'De steekproefophoging bij een gedetailleerde interaktie matrix: enkele problemen en mogelijke oplossingen', in J. van Est, et al. (eds.), *Transportation Planning Research Colloquium*, Delft, 465-498.

Buro Goudappel en Coeffeng (1972), 'Vergelijkend Onderzoek Dataverzameling, Analyse en Ritdistributie', *Huisenquete, rapport 1 en 2*, Deventer.

Est, J. van (1976), *CADSS: concentration and deconcentration simulation study; a spatial model for policy, design and research*, Research Centre for Physical Planning TNO, Delft.

Est, J. van and van Setten, A. (1977), 'Calibration methods: application for some spatial distribution models', *Working Paper 6*, Research Centre for Physical Planning TNO, Delft.

Est, J. van and van Setten, A. (1978), 'Least-squares procedures for a spatial multiplicative distribution model', P. Bovy, et al. (eds.), *Transportation Planning Research Colloquium*, Delft.

Johnston, J. (1972), *Econometric methods*, McGraw-Hill Book, Tokyo.

Nakanishi, M. and Cooper, L. G. (1974), 'Paramter estimation for a multiplicative competitive interaction model – least-squares approach', *Journal of Marketing Research*, 11, 303-311.

Stopher, P. (1975), 'Goodness-of-fit measures for probabilistic travel demand models', *Transportation*, 4, 67-83.

Theil, H. (1971), *Principles of econometrics*, John Wiley, New York.

Williams, I. (1974), 'A comment on the derivation of calibration criteria by Batty and Mackey', *Environment and Planning*, 6, 603-605.

Wilson, A. G. (1974), *Urban and Regional Models in Geography and Planning*, John Wiley, London.

Part 4: The introduction of stochastics in regional control

10. Stochastic control of regional economies

R.J. BENNETT and K.C. TAN*

10.1 INTRODUCTION

There has been considerable and increasing interest in the possibility of applying regional science models to problems in economic policy. The aim of such applications is summarised by Isard (1975, p. 495) as ranging from accurate description of present economic states and trends, to planning of industries, towns and transport facilities and at the most complicated level to 'proposing the structure and functions of a new mode within the hierarchical structure of the world order, capable of evolving and administering a world environmental-management policy and a multiregion development program'. Despite such aims, however, the models used in regional science often show little consideration for the effects of policy in the formulation of their equations; there is often a complete lack of specification of any policy instruments with which to operate policy, and most models are based on static equilibrium assumptions.

It is the aim of this paper to overcome some of these inadequacies using techniques deriving from the mathematical theory of optimal control. Although the theory of optimal control has been developed for physical systems governed by fairly constant parameter equations and spatial relationships with continuously differentiable partial differential operators, it is possible to make ready application of control theory to economic problems of regional science. The development of such models must cope with higher degrees of uncertainty and spatial relationships described by discrete variables. For this reason there have as yet been few applications of optimal control theory to regional science. Indeed most applications of

*R. J. Bennett, Department of Geography, University of Cambridge; K. C. Tan, Department of Geography, University College London.

control theory to economic systems have been based on macroeconomic management in national economies in which it has usually been assumed the system model is known a priori. These applications draw on the mathematical results of Pontryagin et al. (1962) and Kalman (1960), developed for engineering systems. Important macroeconomic applications are Tustin (1953), Theil (1966), Pindyck (1973), Vishwakarma (1974), Chow (1975), Aoki (1976) and Bennett and Chorley (1978).

The present paper develops in section 10.2 a description of national economies disaggregated into N regions for which optimal control algorithms can be developed. Section 10.3 is then concerned with the deriving T-stage control strategies under uncertainty for this system of regions. Section 10.4 deals with the particular mathematical difficulties that arise in multilevel economies which require either joint optimization methods, or optimization subject to constraints. A simple mathematical example is described in this section which relates to allocation of the local authority Rate Support Grant in the U.K. Section 10.5 treats some of the problems that remain to be answered in research for practical control applications. Especially important are the interactions between parameter estimation and control solutions.

10.2 MATHEMATICAL REPRESENTATION OF REGIONAL SYSTEMS

The development of mathematical models of regional systems require models which are truly dynamic in their specification. Most models which represent the dynamics of urban or regional economies have been built on the basis of the Lowry (1964) and Wilson (1974) interaction-allocation framework and are often comparative-static. Other developments have sought to allocate changes to regional distributions of population and economic activity introduced via forecasts of changes in the levels of basic and exogenously determined employment. Other approaches to modelling of regional dynamics have been based on a series of themes, e.g. simulation, econometric and time series models (see Bartels, this volume, and Bennett, 1978). Despite the breadth of interest in these models of regional dynamics, few of these models take explicit account of the needs of policy and it is possible to make four criticisms of these approaches. First, few of the models include specific policy instruments that can be manipulated in order to make policy choices.

Second, none of the models allow for explicit specification of policy goals. Third, few of the models have attempted evaluation or calibration of policy impacts. Some regional econometric models have tried to specify the effects of policy instruments determined a posteriori from the effects of past applications of given policies, but such models are very restricted in applicability in that sufficient time series of data on past policy measures must be available. In addition the policy instruments and impacts must be quantifiable, and the policy application must have covered a similar range of control magnitude to those envisaged in the future (in order to avoid cases in which new control application may affect system stability or go outside of known regimes of behaviour). A final criticism is that few regional models have attempted to make policy choices in an optimal fashion over extended planning horizons.

Most models of regional dynamics have been developed using the structural form (Bennett, 1978):

$$(\boldsymbol{A}_0+\boldsymbol{A}_1 z+\ldots\boldsymbol{A}_p z^p)\boldsymbol{y}_t=(\boldsymbol{B}_0+\boldsymbol{B}_1 z+\ldots\boldsymbol{B}_q z^q)\boldsymbol{u}_t+\boldsymbol{e}_t \qquad (2.1)$$

in which $\boldsymbol{y}_t$ is a $(N\times 1)$ vector of system endogenous variables; $\boldsymbol{u}_t$ is a $(N\times 1)$ vector of system control variables; $\boldsymbol{e}_t$ is a $(N\times 1)$ vector of independent random variables with zero mean and diagonal covariance matrix $\boldsymbol{R}$; $\boldsymbol{A}_i$ and $\boldsymbol{B}_i$ are each $(N\times N)$ matrices of parameters, and z is the unit shift operator: $z\boldsymbol{y}_t = y_{t-1}$. The dimension N of the vectors refers to the N regions of a given national economy such that there is one endogenous and one control variable for each region. This univariate representation is retained for simplicity below, but all results hold equally when the input-output data are multivariate, i.e. there are n inputs and m outputs.

The structural form (2.1) is a particularly useful representation for econometric estimation using time-series data (Bartels, this volume). Although it takes no explicit account of the distance between regions, any distance decay effects are absorbed into the magnitude and sign of the $\{\boldsymbol{A}_i\}$ and $\{\boldsymbol{B}_i\}$ parameters. Hence, the model is not constrained by a particular distance metric and any form of interzonal interaction can be reproduced. The $\boldsymbol{A}_i$ matrices represent autoregressive, or feedback terms and the $\boldsymbol{B}_i$ matrices represent distributed lags of the N outputs $\boldsymbol{y}_t$ on past values of the exogenous or control variable $\boldsymbol{u}_t$. The structural form (2.1) is, however, inconvenient for the development of optimal control models of regional dynamics. A commonly used method (e.g. Chow, 1975; Pindyck,

1973; Turnovsky, 1977) is to transform the model (2.1) into an equivalent first-order system which is given by

$$\boldsymbol{Y}_t = \boldsymbol{A}\boldsymbol{Y}_{t-1} + \boldsymbol{B}\boldsymbol{u}_t + \boldsymbol{v}_t \tag{2.2}$$

where

$$\boldsymbol{Y}_t = (\boldsymbol{y}_t', \ldots, \boldsymbol{y}_{t-p+1}' \,\vdots\, \boldsymbol{u}_t', \ldots, \boldsymbol{u}_{t-q+1}')'$$

$$\boldsymbol{v}_t = \begin{bmatrix} A_o^{-1} e_t \\ 0 \\ \vdots \\ 0 \\ 0 \\ \vdots \\ 0 \end{bmatrix}, \quad \boldsymbol{B} = \begin{bmatrix} A_o^{-1} B_o \\ 0 \\ \vdots \\ 0 \\ I \\ 0 \\ \vdots \\ 0 \end{bmatrix}$$

$$\boldsymbol{A} = \begin{bmatrix} -A_o^{-1}A_1 & -A_o^{-1}A_2 & \ldots\ldots & -A_o^{-1}A_p & A_o^{-1}B_1 & \ldots\ldots & A_o^{-1}B_q \\ I & 0 & \ldots\ldots & 0 & 0 & \ldots\ldots & 0 \\ \vdots & & & \vdots & \vdots & & \\ 0 & 0 & \ldots\ldots I & 0 & 0 & \ldots\ldots & 0 \\ 0 & 0 & \ldots\ldots & 0 & 0 & \ldots\ldots & 0 \\ \vdots & & & \vdots & I & \ldots\ldots & 0 \\ \vdots & & & \vdots & \vdots & & \\ 0 & 0 & \ldots\ldots & 0 & 0 & \ldots\ldots I & 0 \end{bmatrix} \tag{2.3}$$

and $\boldsymbol{Y}_t, \boldsymbol{v}_t$ are vectors of dimension $(p+q)N \times 1$.

This is a state-space form representation since the values of the combined output and input (control) vector $\boldsymbol{Y}_t$ now depend only on the previous values $\boldsymbol{Y}_{t-1}$, and the setting of $\boldsymbol{u}_t$ can be chosen arbitrarily. There are several reasons why (2.2) is preferred. First, it simplifies the algebraic manipulations when the optimal control solution is derived. Second, the usual quadratic objective function used by control engineers with separate costs of deviations from desired values for the outputs $\boldsymbol{y}_t$ and the controls $\boldsymbol{u}_t$ is simplified to one where the argument of the objective function is only in terms of $\boldsymbol{Y}_t$. Hence it is not necessary to assume that the weighting matrix for deviations of the controls from desired values be non-singular (Pindyck, 1973; Chow, 1975). Third, the transformation (2.2) preserves the structure of the underlying model and allows the optimal control solutions to be easily interpreted (Turnovsky, 1977). It is well-known that the above state-space representation is not unique (Preston, 1977) and is not the usual state-space representation familiar to control theorists. The lack of uniqueness means that, in general, there exists an infinite set of realisations $\{\boldsymbol{A},\boldsymbol{B}\}$ which will satisfy (2.2). It is usual in control theory to restrict attention to those realisations which give the smallest possible state space representation, and such realisations are termed *minimal realisations*. Minimal realisations have the important property that the system equation satisfies two conditions of observability and controllability. These conditions assure that the minimal realisation of $\{\boldsymbol{A},\boldsymbol{B}\}$ satisfies conditions of rank, and that there are no 'self cancelling' modes of behaviour which limit the capacity to measure or to act on any element in the system equation. A realisation is minimal if and only if it is controllable and observable (Kalman, 1960). In general, (2.2) is not minimal. It is possible to determine a minimal realisation of the system structure from (2.1), and this will have many attractive mathematical properties, especially of controllability and observability. But for regional economic systems minimal realisations are not usually very easily derived, nor very attractive. It becomes extremely difficult to interpret the various variables, and the minimal form equations may give a very inconvenient structure which makes it of little utility for practical implementation of regional economic policies (Caines and Wall, 1972). Hence, for all developments below, the structure of (2.2) is retained, whilst recognising that, in some instances, this will not constitute a minimal realisation.

It is convenient for parameter estimation that the matrix $\boldsymbol{A}_o$ is diagonal. This allows the structure of the disturbance term in (2.1) to be maintained in (2.2) except for the addition of a constant factor. This assumption is not necessary for control purposes where, in general, $\boldsymbol{v}_t$ may be allowed to become a vector of correlated random variables with zero mean and covariance matrix given by

$$\begin{aligned}\text{Cov}(\boldsymbol{v}_t) &= \text{Cov}(\boldsymbol{A}_o^{-1}\boldsymbol{e}_t)\\ &= \text{E}(\boldsymbol{A}_0^{-1}\boldsymbol{e}_t\boldsymbol{e}_t'\boldsymbol{A}_o^{-1\,\prime})\\ &= \boldsymbol{A}_o^{-1}\boldsymbol{R}\boldsymbol{A}_o^{-1\,\prime}\\ &= \boldsymbol{R}^* \end{aligned} \tag{2.4}$$

However it has been usual to assume that $\boldsymbol{A}_o$ is an identity matrix (see for example Chow, 1975, and Turnovsky, 1977), but in the developments below the definition (2.4) is used.

10.3 OPTIMAL CONTROL MODELS OF REGIONAL SYSTEMS

The basic aim of control models is to manipulate a set of policy variables (instruments) in order to achieve a set of goals or targets. The targets are specified in terms of desired levels of the dependent variable vector $\boldsymbol{y}_t$. Hence target levels of $\boldsymbol{Y}_t$ determine the level of instruments $\boldsymbol{u}_t$ and causality, in the resulting economic model, is reversed. This approach can be distinguished from that used by econometricians where simulation of the effects of various levels of $\boldsymbol{u}_t$ is used to choose a tolerable level of $\boldsymbol{Y}_t$. The econometric approach is more explanatory and is less technocratic, but it does not generally yield optimal results.

For control of regional systems the aim of control models is to determine the optimal settings of the policy instruments in each region to achieve a set of different targets in each region. Most existing developments of optimal control theory for spatially distributed systems are based on models described by partial differential and difference equations (see for example Butkovskii, 1969). For such models it must be assumed that the space is continuously differentiable or at least regularly partitioned (e.g. on a Cartesian lattice) and that interaction between locations is controlled by a pattern of distance decay. In addition it is usually assumed that at least the structure (but not necessarily the parameters) of the model is known a priori. Such models have been developed mostly in engineering for examples such as control of heat flux in kilns and control of chemical reactions. Spatial models described by partial differential equations are of little utility for developing control models of regional economies, and this paper is concerned with extending the results of optimal control theory to situations in which the spatial domain of interest (e.g. a national economy) is partitioned into N regions which interact arbitrarily such that no distance metric is necessarily involved.

Four possible approaches to the development of control models of regional economies have been recognised by Bennett and Chorley (1978). First, *dynamic control* is concerned with regional economies described by (2.2), but which are aggregated to an overall lumped form as in national macroeconomic planning. Second, *spatial control* is concerned with regional economies which are disaggregated but in which no account is taken of temporal evolution. This latter approach characterises many decisions on the optimal layout of service facilities (e.g. schools, hospitals) and also electoral and administrative districting. A third approach is by *separation of space and time variables* in which regional economic control with (2.2) is separated into two functions, one depending only on time, and the other depending only on space (Butkovskii, 1969). Fourier expansion and the Galerkin technique are examples of this method (see Polis, Gondson and Wozny, 1973). The final approach is concerned with the development of *complete regulatory and servomechanism control* solutions in which full account is taken of spatial and temporal dependence, and in which redesign and adaptation of system coefficients can also be included. The regulation problem refers to the case in which it is desireable to keep a regional system at a particular target or set point, whilst the servomechanism problem refers to the case in which it is desireable to make the regional system conform to a trajectory of targets which change over time. It is clear that the first three approaches provide only approximate solutions appropriate to special cases and only the regulation and servomechanism approaches are developed below.

For general control problems involving the regulator or servomechanism it is necessary to weight deviations from targets by a criterion of error or loss. This represents the introduction of an objective function on the part of the policy maker in order to represent the relative importance of large versus small deviations, or to weight differentially deviations with respect to some variables or regions relative to others. Any criterion of error can be used, but a frequently used criterion which has many useful mathematical properties is the quadratic objective function which is adopted below:

$$\mathrm{E}(W) = \mathrm{E} \sum_{t=1}^{T} [(\boldsymbol{Y}_t - \boldsymbol{r}_t)'\boldsymbol{M}(\boldsymbol{Y}_t - \boldsymbol{r}_t)] \tag{3.1}$$

The $\mathrm{E}(W)$ term is the expectation of the objective function in which $\boldsymbol{r}_t$ is a $(p+q)N \times 1$ vector of targets and $\boldsymbol{M}$ is a $(p+q)N \times (p+q)N$ symmetric positive semi-definite matrix of weights which is assumed *known*. When $\boldsymbol{r}_t = \boldsymbol{r}$ for all t, i.e. a constant, we have a regulator problem, and when $\boldsymbol{r}_t$ is a changing set of targets there is a regional servomechanism problem. When $T = 1$ we have a one-stage control problem. But in regional economic control we must almost always take $T > 1$, first, because regional economies are sluggish, the response to control is usually slow; industrial inertia and the length of learning time required in social systems limits the magnitude and rapidity of changes such that control is not socially or politically disruptive. Secondly, regional economic control is cost-limited in that the financial resources available are usually insufficient to achieve a one-term controller given the marginal return to control and the opportunity costs on capital. Thirdly, our understanding of most regional economies is limited to broad ranges of behaviour which can be predicted with reasonable accuracy and it may not be possible to apply the magnitude of control input necessary to achieve the desired final state in one step. This limits the choice of controls to an admissable set determined by stability and other criteria.

Given the model of regional economies (2.2) and initial conditions $\boldsymbol{Y}_o$, the optimal control problem is the choose that set of N spatial and T temporal controls $\boldsymbol{u}_1, \ldots, \boldsymbol{u}_T$ to minimise the welfare function (3.1). That is, we require to choose those values of the policy instruments over the N regions and T time periods so as to render deviations from target levels of $\boldsymbol{Y}_t$ a minimum. In the present case we seek to weight the deviations by the quadratic criterion (3.1). The regional model (2.2) is identical in structure to multivariate models of lumped systems. Using this symmetry property, we may write the optimal control solution directly using the results of the Kalman (1960) theory for the linear quadratic control problem. For economic systems Chow (1975) gives a useful expression of this control solution as

$$u_t^o = -(B'H_tB)^{-1}B'H_tAY_{t-1} + (B'H_tB)^{-1}B'h_t \tag{3.2}$$

Equation (3.2) is a recursive structure which yields the optimal settings of the control variable at each time. The set of matrix terms premultiplying Y_{t-1} represent the Kalman gains K and the second set of terms derive from the target settings. H_t has dimension $(p+q)N \times (p+q)N$, and h_t is a $(p+q)N \times 1$ vector. These two terms can be derived by a number of mathematical methods, dynamic programming, calculus of variations, and Lagrange multipliers being the most common. Using dynamic programming the standard result is a set of matrix Ricatti equations given by

$$\begin{aligned} H_{t-1} &= M + A'H_t(A - B(B'H_tB)^{-1}B'H_tA) \\ H_T &= M \end{aligned} \tag{3.3}$$

and

$$\begin{aligned} h_{t-1} &= Mr_{t-1} + A'h_t - A'H_tB(B'H_tB)^{-1}B'h_t \\ h_T &= Mr_T \end{aligned} \tag{3.4}$$

Provided that we specify the data on the initial conditions Y_o, we may integrate (3.3) and (3.4) forwards in time from these specified initial conditions to determine the optimal control setting u_t^o at each intermediate time point over the T stages of control. The facility with which optimal quadratic controllers for regional economies can be developed in the form (3.2) to (3.4) above exposes a number of outstanding research problems which are discussed in the next section.

It should be noted that although the representation (2.2) upon which (3.2) is based is not minimal, the control solution will yield an optimal solution provided that there are at least the same number of targets as there are instruments, and that the number of targets is equal to the number of regional outputs.

A simple example of a two-region economy will exhibit the use of (3.2) to (3.4). Consider a model (2.2) of a two-region economy in reduced form and, described by

$$\begin{bmatrix} Y_{t,1} \\ Y_{t,2} \end{bmatrix} = \begin{bmatrix} 2 & 1 \\ 0 & 2 \end{bmatrix} \begin{bmatrix} Y_{t-1,2} \\ Y_{t-1,2} \end{bmatrix} + \begin{bmatrix} 1 & 0 \\ 2 & 1 \end{bmatrix} \begin{bmatrix} u_{t,1} \\ u_{t,2} \end{bmatrix} + \begin{bmatrix} v_{t,1} \\ v_{t,2} \end{bmatrix} \tag{3.5}$$

It is desired to move the endogenous variables $\boldsymbol{Y}_t$ from its initial values to achieve the target setting $\boldsymbol{r}_t$ according to the optimal control equation (3.2) using quadratic costs (3.1). As a simple example we assume that we have a two stage problem, i.e. $T = 2$. If we let

$$\begin{aligned} \boldsymbol{Y}_o &= (3,-1)', \boldsymbol{r}_1 = (5,3)', \boldsymbol{r}_2 = (8,6)', \boldsymbol{v}_1 = \boldsymbol{v}_2 = (0,0)' \\ \boldsymbol{M} &= \boldsymbol{I}, \text{ the identity matrix} \end{aligned} \tag{3.6}$$

the the optimal control solutions are found to be

$$\begin{bmatrix} u^o_{1,1} \\ u^o_{1,2} \end{bmatrix} = \begin{bmatrix} 0 \\ 5 \end{bmatrix} ; \begin{bmatrix} u^o_{2,1} \\ u^o_{2,2} \end{bmatrix} = \begin{bmatrix} -5 \\ 10 \end{bmatrix} \tag{3.7}$$

For simplicity of calculation the error term in (3.5) has been set to zero such that in this example the targets can be achieved exactly at each stage. This result is expected since with the elimination of the error terms, the problem has become a deterministic problem and since also the number of targets equals the number of instruments; it is also controllable and observable. The sort of examples in which models such as (3.5) arise are in setting optimal levels of financial subsidies to sets of regions to achieve economic, social or expenditure targets for levels of Y_t. Further examples are discussed in the next section where it is shown that such policies usually involve the interaction of local and central levels of government.

All along we have assumed that the inverse of the matrix $(\boldsymbol{B}'\boldsymbol{H}_t\boldsymbol{B})$ in (3.2) exists. It can be seen that the inverse will exist provided the number of control variables is less than or equal to the number of target variables. Necessary and sufficient conditions for the existence of the inverse are given by Garbade (1976).

Taking (3.5) as a deterministic model, it can be easily verified that the Kalman (1960) controllability and observability conditions are satisfied. In fact for the deterministic equivalent system given by (2.2), it is always observable as the state and observable output are the same thing. Regarding the controllability property, when the controls are constrained as discussed in section 4, it is in general true that the system (2.2) is uncontrollable. However, optimal control solutions for the constrained problem do exist and are subject to the same existence conditions given by Garbade (see Tan, 1978a). In other words, for the control problem with constrained controls, it is more meaningful to consider the existence (and uniqueness) of optimal solutions rather than the controllability of the system.

Control models in the form of the Kalman structure have been given by Nijkamp (1975) and Nijkamp and Verhage (1976) for control of pollution problems, and for regional economic problems by Domanski (1973). Other important cases in which regional control models may be developed are the disposition of regional policy instruments within a national economy (e.g. the determination of the optimal levels of labour subsidies used as control instruments to achieve given employment targets in each of the N regions) and the setting levels for control of international capital plans, trade and aid to achieve given equity or development targets in different regions.

10.4 INTERACTION OF OPTIMAL CONTROL OF REGIONAL ECONOMIES WITH NATIONAL GOVERNMENTS

The control rule (3.2) and system model (2.2) assume that each of the N regions interacts in an arbitrary way. However, frequently the interaction structure is constrained in a number of ways. A constraint frequently introduced by geographers is that of a distance relation between zones. Other important constraints enter when there are hierarchical effects which govern interaction between regions, for example, urban hierarchies in central place systems. The particular set of constraints discussed here are those which arise when an overall government (usually a national government) interacts with a regional level. In federal systems and in centralised states with devolved agencies there are great difficulties in reconciling regional control optima with national control objectives. Practical attempts to integrate these various levels have evidenced the difficulties involved (see Klaassen and Paelinck, 1974). Two main mathematical approaches are possible based upon either multilevel control or the determination of optima under constraints.

1. *Multilevel control* treats the integration problem as a hierarchical joint optimisation of both regional and national control targets which creates a number of difficulties in practice. The joint optimisation problem is usually of large dimension and there will be the frequent appearance of near-equivalent local optima. One approach to the dimensionality problem in practice, of course, is that many economic systems employ decentralisation to federal or devolved decision units. A number of workers have considered control of systems which are completely decentralised, i.e. decoupled (see Schoeffler, 1971). When the regional system is not constitutionally decoupled it is necessary to adopt a number of mathematical devices to reduce the system dimensions. Three forms of such mathematical decomposition can be adopted (Schoeffler, 1971; Bennett and Chorley, 1978). The first, structural decomposition, involves partitioning the system into separate subsets, one for each region, each with independent targets and controls which are then coordinated. A second method, influence decomposition, partitions the controls into levels based upon which levels set preemptive target values for the others, i.e. splitting the economy into executive (policy-determining) and subordinate regions. Control decomposition provides a third method which uses any of the special structural properties of the mathematics of the control solution itself to simplify to independent optimisation problems. The advantages and disadvantages of the decomposition approaches are well illustrated by the difficulties of coordinating the controls in the decomposed models. One *dual-feasible* method given by Schoeffler (1971) uses a disaggregation of the objective function (3.1) into N parts, one for each region. The interaction between the N regions is taken into a set of ($N \times N$) slack variables $\boldsymbol{Z}_t$ which allow complete decoupling with the real interaction effects set equal to the slack variables. For the resulting two levels the control problem at level 1 is solved for the decoupled set of N independent systems. Then at level 2, the first level solution is forced to satisfy an interaction-balance principle. The difference between the true system interactions $\boldsymbol{X}_t$ and the slack variables $\boldsymbol{Z}_t$ is used to define a set of Lagrangian multipliers λ for which a necessary condition for a minimum is determined when the derivative equals zero. Hence for a two-region system, optimisation is carried out over the modified cost function:

$$\mathrm{E}(W) = \mathrm{E}(W_1(U_{t1},X_{t1})) + \mathrm{E}(W_2(U_{t2},X_{t2})) + \boldsymbol{\lambda}'(X_t - Z_t)$$

In practice this optimum must often be determined iteratively. The multilevel approach results in problems with large dimensionality. It may often be preferable, therefore, to use a second approach developed below.

2. The *introduction of constraints* often allows many of the dimensionality problems of multilevel control to be overcome by designing control systems incorporating specific constraints on the interaction of ther various levels of economic management. The use of constraints to attain this goal has been suggested by Nijkamp (1975) but he considers constraints only on the state variables Y_t. Mathematical results for problems involving constraints on the control variables rely on the results of Tan (1978a, 1978b). In the case of linear equality constraints, the optimal control problem is formulated as: given the model (2.2) with the initial condition $\boldsymbol{Y}_o$ the optimal control problem is to find a new set of control variables $\boldsymbol{u}_1,\ldots,\boldsymbol{u}_T$ to minimise the quadratic objective function (3.1) when the control variables $\boldsymbol{u}_t$ are subject to a set of $w < N$ independent linear constraints

$$\boldsymbol{F}_t\boldsymbol{u}_t = \boldsymbol{f}_t\ ,\ t = 1, \ldots, T \tag{4.1}$$

in which $\boldsymbol{f}_t$ is a $(w \times 1)$ vector of specified central government spending targets and $\boldsymbol{F}_t$ is a $(w \times N)$ matrix of known constants. The optimal control rule is then given, following Tan (1978a), by

$$\boldsymbol{u}_t^o = \hat{\boldsymbol{G}}_t \boldsymbol{Y}_{t-1} + \hat{\boldsymbol{g}}_t \tag{4.2}$$

where $\hat{\boldsymbol{G}}_t$ is of dimension $(N \times (p+q)N)$, and where

$$\hat{\boldsymbol{G}}_t = \boldsymbol{G}_t - \boldsymbol{D}_t^{-1} \boldsymbol{F}_t' \boldsymbol{P}_t^{-1} \boldsymbol{F}_t' \boldsymbol{G}_t \tag{4.3}$$

$$\hat{\boldsymbol{g}}_t = \boldsymbol{g}_t - \boldsymbol{D}_t^{-1} \boldsymbol{F}_t' \boldsymbol{P}_t^{-1} \boldsymbol{F}_t \boldsymbol{g}_t + \boldsymbol{D}_t^{-1} \boldsymbol{F}_t' \boldsymbol{P}_t^{-1} \boldsymbol{f}_t \tag{4.4}$$

$$\hat{\boldsymbol{G}}_t = -\boldsymbol{D}_t^{-1} \boldsymbol{B}' \boldsymbol{H}_t \boldsymbol{A} \tag{4.5}$$

$$g_t = D_t^{-1} B' h_t \tag{4.6}$$

$$D_t = B' H_t B \tag{4.7}$$

$$P_t = F_t D_t^{-1} F_t' \tag{4.8}$$

which is a modified form of (3.2) with H_t and h_t satisfying the new set of matrix Ricatti equations given by

$$H_{t-1} = M + A' H_{t-1} (A + B\hat{G}_t) \tag{4.9}$$

$$H_T = M$$

and

$$h_{t-1} = Mr_{t-1} + A' (h_t - H_t B\hat{g}_t) \tag{4.10}$$

$$h_T = Mr_T$$

Equation (4.2) is identical in the deterministic and stochastic cases, i.e. certainty-equivalence applies when the controls are subject to linear equality constraints.

In the case of linear inequality constraints an optimal control rule is required, subject to the addition of the constraint

$$u_t \geqslant 0, \quad t = 1, \ldots, T \tag{4.11}$$

to the control problem constraints (4.1). As shown by Tan (1978b) it is not possible in this case to obtain analytic solutions. An alternative strategy proposed by Tan may be used to give the optimal solutions and this is composed of two parts. First, solve the control problem with only (4.1) present. Second, if the resulting solution is not feasible (i.e. there exists at least one t such that $u_t^o \geqslant 0$), then adjust $u_1^o, \ldots, u_T^o$ simultaneously. The adjustment mechanism is based on the dual simplex method (Hadley, 1962). Making the definition

$$U = (u_1', \ldots, u_T')' \tag{4.12}$$

the optimal controls U^* for the problem with constraints (4.11) added must satisfy the following conditions:

$$U^* = \hat{G}_s Y_o + \hat{g}_s + \bar{B}_s \beta^* \geqslant 0 \tag{4.13}$$

$$U^* \geqslant 0 ; \beta^* \geqslant 0 \text{ and } \beta^{*\prime} U^* = 0 \tag{4.14}$$

where β^* is the optimal Lagrange multiplier. In (4.13) the quantities $\hat{G}_s$, $\hat{g}_s$ and $\bar{B}_s$ are given by

$$\hat{G}_s = G_s - B_s^{-1} F' P^{-1} F G_s \tag{4.15}$$

$$\hat{g}_s = g_s - B_s^{-1} F' P^{-1} F g_s + B^{-1} F' P^{-1} f \tag{4.16}$$

$$\bar{B}_s = B_s^{-1} - B_s^{-1} F' P^{-1} F B_s^{-1} \tag{4.17}$$

$$G_s = -B_s^{-1} A_s \ ; \ g_s = B_s^{-1} h \ ; \ P = F B_s^{-1} F' \tag{4.18}$$

where

F = diag $(F_1, \ldots, F_T)$ a block diagonal matrix of dimension $Tw \times TN$

f = $(f_1', \ldots, f_T')'$ a $Tw \times 1$ vector

$$A_s = \begin{bmatrix} B'MA + B'A'MA^2 + \cdots + B'A'^{T-1}MA^T \\ \vdots \\ B'MA^{T-1} + B'A'MA^T \\ B'MA^T \end{bmatrix}$$

which has dimension $TN \times (p+q)N$

$$h = \begin{bmatrix} B'Mr_1 + B'A'Mr_2 + \cdots + B'A'^{T-1}Mr_T \\ \vdots \\ B'Mr_{T-1} + B'A'Mr_T \\ B'Mr_T \end{bmatrix}$$

which is of dimension $TN \times 1$, and $\boldsymbol{B}_s$ is a symmetric $(TN \times TN)$ matrix, the elements of which are given by $\boldsymbol{B}_s = (b_{ij})$ $i, j = 1, \ldots, T$ where

$$b_{ii} = \boldsymbol{B}' \, \boldsymbol{C}_i \boldsymbol{B}$$

$$b_{ij} = \boldsymbol{B}' \, \boldsymbol{C}_i \boldsymbol{A}^{i-j} \boldsymbol{B} \quad \text{for } i > j \, .$$

The $\boldsymbol{C}_i'$'s are found from the following equations:

$$\boldsymbol{C}_T = \boldsymbol{M}$$

$$\boldsymbol{C}_{i-1} = \boldsymbol{M} + \boldsymbol{A}' \, \boldsymbol{C}_i \boldsymbol{A} \quad i = 1, \ldots, T$$

Observe that when the constraints (4.11) are absent then (4.13) reduces to

$$\boldsymbol{U}^o = (\boldsymbol{u}_1^{o\,\prime}, \ldots, \boldsymbol{u}_T^{o\,\prime})' = \hat{\boldsymbol{G}}_s \, \boldsymbol{Y}_o + \hat{\boldsymbol{g}}_s = \overline{\boldsymbol{G}}_s \tag{4.19}$$

which is exactly the optimal control solutions with equality constraints (4.1) only. Given (4.19) which does not satisfy (4.14), the second step of the strategy considers the following programming problem:

$$\text{Max } W = \boldsymbol{\iota}_v' \; (\boldsymbol{FU}\text{-}\boldsymbol{f}) \text{ subject to} \tag{4.20}$$

$$\boldsymbol{U}\text{-}\overline{\boldsymbol{B}}_s \boldsymbol{\beta} = \overline{\boldsymbol{G}}_s$$

$$\boldsymbol{U} \geqslant \boldsymbol{0}$$

$$\boldsymbol{\beta} \geqslant \boldsymbol{0}$$

$$\boldsymbol{\beta}' \boldsymbol{U} = \boldsymbol{0}$$

where $\boldsymbol{\iota}_v'$ is a $(Tw \times 1)$ vector whose elements are all unity. It is easily checked (Tan, 1978b) that the initial basic solution $\boldsymbol{U}_b$ given by (4.19) is primal infeasible, but is dual feasible. Hence the dual simplex algorithm may be employed to determine the optimal values $\boldsymbol{U}^*$ and $\boldsymbol{\beta}^*$. However, because of the constraint $\boldsymbol{\beta}' \boldsymbol{U} = \boldsymbol{0}$, a variation to the algorithm is needed. If $U_i > 0$ then β_i is not allowed to enter the basis, and vice versa. This variation is exactly similar to that found in Wolfe's (1959) algorithm for solving quadratic programming problems. Finally, at each iteration of the algorithm the objective function W (equation 4.20) is always at the zero level. For well-defined problems, the algorithm will terminate to an optimal solution in a finite number of steps.

The above strategy gives a truly optimal solution to systems which are deterministic. When an additive vector of random disturbances is present, as in (2.2), Tan (1978b) points out that the certainty equivalence principle will not be satisfied and will not allow a truly optimal solution. It is already well known (Chow, 1975) that this principle breaks down for nonlinear systems without constraints on controls. It would then seem fair to suggest that whenever there is any nonlinearity in either the known model or the objective function or constraints on controls and states, certainty equivalence is not applicable to giving optimal solutions.

An important example in which the constrained solutions discussed here arise is in the allocation of financial resources from a central government to regions or local authorities. In the UK the allocation of the Rate Support Grant to local authorities represents an important class of such problems. In this case the control variable is a financial or fiscal instrument and targets are specified in terms of social indicators (such as social service, health, education and welfare needs) in each region. The flow of resources required is from central to local areas and in such cases it is often necessary to impose an equality constraint such that total central government expenditures do not exceed a predetermined figure. Inequality constraints enter because it is usually politically necessary to prevent inequalities in local tax bases from being used to precept taxation income away from better-off regions to give to other regions, i.e. all equalisation between regional expenditure needs and the necessary financial resources is achieved by allocation from central government; regions with an excess of local revenue will have control flows set to zero. Thus it is necessary to impose constraints on the control variables to ensure that they are non-negative as in equation (4.11).

To see how the above solutions of the control problem with constraints is used in the allocation of the Rate Support Grant we can, without loss of generality, consider the determination of the "needs element" which is one of the three constituents that make up the

Rate Support Grant. The state variables $\boldsymbol{Y}_t$ are socio-economic indicators for the N regions and the control vector $\boldsymbol{u}_t$ denotes the flow of grants to the N regions. The present practice of determining the needs element is based on a multiple regression formula. For each region the dependent variable is the cash grant to be received and the independent variables are the socio-economic indicators. This multiple regression form can be translated into the reduced form (2.2), the details of which are given by Bennett and Tan (1978).

As the total of grants distributed each year is considerable, it is important that the distribution is equitable among regions, and hence an optimal method of allocation is desirable. Furthermore, this distribution decision should be connected to an overall central plan (e.g. 5-year economic plans under PESC). A common way in which this feature is achieved is that the central authority stipulates the amount of money it is each year prepared to distribute between the N regions. This specification by the central authority can be represented by the constraint (4.1) where $\boldsymbol{F}_t$ is a row vector whose elements are all unity and f_t is the total amount available for distribution at time t. For reasons of a political nature, the cash grant is allowed to flow only from the central government to the regions and not vice versa. This requirement is achieved by adding the constraints (4.11) to the control problem.

We shall now consider a simple numerical example which captures most of the relevant features of allocation of Rate Support Grant and illustrates the calculations involved in obtaining the optimal control solutions for the case when the control variables are constrained. The numerical example considered in section 10.3 is now considered to be subject to the following equality constraints:

$$\boldsymbol{F}_1 = \boldsymbol{F}_2 = [1 \quad 1] \text{ and } f_1 = 5, f_2 = 8 \qquad (4.21)$$

In considering this example, it is useful to compare the different optimal control solutions under the three conditions:

1. when the control variables are unconstrained, i.e. *basic problem*;

2. when the control variables are subject to linear equality constraints, i.e. *modified problem I*;

3. when the control variables are subject to linear equality con-

straints in (2) as well as the non-negativity restrictions, i.e. *modified problem II.*

The solution to each of the three problems may be summarised as

	Optimal U_t at $t = 1$	Optimal U_t at $t = 2$
Basic problem	$\begin{bmatrix} 0 \\ 5 \end{bmatrix}$	$\begin{bmatrix} -5 \\ 10 \end{bmatrix}$
Modified problem I	$\begin{bmatrix} 0.6 \\ 4.4 \end{bmatrix}$	$\begin{bmatrix} -8 \\ 16 \end{bmatrix}$
Modified problem II	$\begin{bmatrix} 0 \\ 5 \end{bmatrix}$	$\begin{bmatrix} 0 \\ 8 \end{bmatrix}$

This table shows that the introduction of constraints gives significantly different settings of the flow of financial resources under the Rate Support Grant needs element than the unconstrained solution. The results for the inequality constraint case (modified problem II) should be particularly noted. Clearly it is necessary to adjust the solutions of modified problem I in order to incorporate the inequality constraint. The control settings $\boldsymbol{u}_1^o = (0.6\ \ 4.4)'$ and $\boldsymbol{u}_2^o$ do not satisfy the inequality constraint. Both $\boldsymbol{u}_1^o$ and $\boldsymbol{u}_2^o$ must be adjusted simultaneously; it is not sufficient to adjust $\boldsymbol{u}_2^o = (-8\ \ 16)'$ only. It can be easily checked that

$$\bar{\boldsymbol{B}}_s = 1/5 \begin{bmatrix} 2 & -2 & -5 & 5 \\ -2 & 2 & 5 & -5 \\ -5 & 5 & 15 & -15 \\ 5 & -5 & -15 & 15 \end{bmatrix}$$

and that the dual simplex algorithm terminates after two iterations giving the optimal solutions

$$U^* = (0\ \ 5\ \ \vdots\ \ 0\ \ 8)'\ ,\ \ \beta^* = (31\ \ 0\ \ \vdots\ \ 13\ \ 0)'$$

The initial primal basic infeasible solution is $U_b = (0.6\ 4.4\ {-8}\ 16)'$.

10.5 PROBLEMS IN APPLYING OPTIMAL CONTROL TO REGIONAL SYSTEMS

The quadratic control solutions discussed above are extremely attractive for use in regional economic control, not least because all of the classical results of mathematical control theory for multivariate systems will apply directly. But before they can be widely utilised, it will be necessary in most instances to overcome a number of major problems. The most important of these, raised by Bennett (1977a), are discussed below.

One major problem concerns the use of the quadratic cost criterion (3.1). This has the advantage of simplicity and of permitting simple mathematical manipulation. However, these advantages should not blind us to the difficulties associated with assuming quadratic costs, i.e. the arbitrariness in treating both positive and negative deviations from targets with equal weight. For example, increases or decreases from a target level of employment, or from a target for satisfaction of social welfare needs, are not equally tolerable. Moreover we would not expect positive and negative deviations to be equally tolerable over all regions; some regions (the more developed) might be able to tolerate positive deviations from employment levels (reducing over-heating) but might be less able to tolerate negative deviations (which generate excessive wage-push inflation). In less developed regions characterised by predominant patterns of labour surplus, the tolerance would be reversed. Long-term research is required in order to determine with greater precision spatial welfare functions alternative to the quadratic, although Livesey (1973) questions whether more complicated welfare functions will be any less arbitrary or more determinate.

A second problem concerns the fact that the system model (2.2) is usually not known with certainty but is derived from parameter estimates using time series or other data. The important consequence is that a separation principle usually does not hold between parameter estimation and control (Åstrom and Wittenmark, 1971). Attempts at separate treatment of the two problems will not yield a global minimum of the objective function (3.1) (see Bennett and Chorley, 1978). In response to this problem we may adopt the Åstrom and Wittenmark 'self-tuning regulator' which gives recursive estimates of the parameters and optimal control at each step, one step at a time, as a special form of Kalman filter. Although this solution gives optimal estimation and control asymptotically, it has not been found possible to extend the results to multivariate or regional systems. In the multivariate and regional cases we are thrown back on sub-optimal solutions, the most useful of which is the learning algorithm given by Chow (1975, chapter 11). Further developments of this method for regional systems are still required.

Related to the separation problem is a third difficulty of identification in a 'closed loop'. When perfect feedback policies have been implemented in a particular regional system, then the natural dynamics of the system itself become cancelled out and are completely unobservable. When sub-optimal or partial control policies are implemented, then natural regional dynamics become partially unobservable. The problem is that, by keeping a regional system at a set point target, exogenous inputs are manipulated as policy instruments in order that the output be unvarying from the target. If this control is perfect, there will be no variation in regional economic indicators at all, and by using only time series of data on system inputs and outputs we will not be able to determine the character of the regional system at all.

A fourth problem is how to make allowance for changes in coefficients and structure of the system model (2.2). Most economic and regional systems are dominated by evolution and shifts in coefficients in their describing models. There has been growing awareness of this problem (Chow, 1975), and it is relatively simple to extend the structure of the control solutions given above to systems with parameter dynamics. What is lacking as yet is a satisfactory method of identifying the parameter models and learning that occurs when these are not known a priori. Evaluation of a number of the available methods (Bennett, 1975, 1977b) does not give unequivocal support to the universal efficiency of any of the presently available techniques. Some Kalman filter estimation methods may perform very adequately in specific instances, but the methods which will be of most utility in any specific case cannot usually be predicted in advance. Again, when both the system model and any model of change in the coefficients are both unknown,the self-tuning regulator Åstrom and Wittenmark (1971) or Chow (1975) learning method must be modified in regional economic control systems. These problems are being currently investigated by the authors.

The final outstanding problem of the control approach taken in (2.2) and (3.2) is that it is well known that regional economies are usually open systems with many cross-flows between regions arising from high levels of migration, journey-to-work, and trade and product flows. This makes the system model complex and undermines the assumption of (2.2) that the control instruments are independent of each other between regions. Dependence enters from the constitutional structure of regional economic systems (as distinct from separate nation-states) which limits the degrees of freedom of economic control even in federal situations. If we require to specify goals independently for each region, new ways of thinking with regard to economic regulators will be required. In almost all decentralised states macroeconomic regulation is maintained as a central government function, but we know from the controllability conditions governing (3.2) that we cannot achieve optimal regional targets without independent control instruments. We are drawn to accept, therefore, that some measure of disaggregation of macroeconomic regulators is required if we are to maintain control of decentralised economic systems. In this case the constrained control solutions developed in the previous section assume a greater importance.

10.6 CONCLUSION

There has been little application of the methods of optimal control theory to the management of regional economic systems. This is surprising since, as briefly sketched in sections 10.1 and 10.2, all of the results of the mathematical control theory of multivariate systems apply, provided that we do not attempt to impose any distance metric on the system. In section 10.3 these results were extended directly to regional systems. This control approach has three advantages. First, it permits consistent allocations over long time horizons. Second, it emphasises goals and poses the question: to achieve a given target set, what instrument settings must be made? Third, it is possible in further developments explicitly to incorporate the effects of uncertainty in the measurement of the socio-economic targets and models. In section 4 the methods of regional control are extended to systems in which hierarchical equality and inequality constraints are imposed on the control variables. One example of local authority finance in the UK evidences the usefulness of this method. In the final section 10.5 a number of important research problems were discussed. These problems concern the way in which we design control systems to satisfy controllability criteria, and the need for developments of regional control theory which can handle uncertain models with uncertain adaptive coefficients having linear relations between the control instruments. Despite these research problems, it is now possible to develop optimal control solutions for regional economies, and these methods are now being applied by the authors to the full scale problem of allocation of Rate Support Grant in the UK.

REFERENCES

Aoki, M. (1976), *Optimal control and system theory in dynamic economic analysis*, North Holland, Amsterdam.

Åstrom, K. J. and Wittenmark, B. (1971), 'Problems of identification and control', *Journal of Mathematical Analysis and its Applications*, 34, 90–113.

Bennett, R. J. (1975), 'Nonstationary parameter estimation for small sample situations: A comparison of methods', *International Journal of Systems Science*, 7, 257-275.

Bennett, R. J. (1977a), 'Optimal control models of regional economies', *IFAC Workshop on urban, regional and national planning: environmental impacts, Kyoto*, Pergamon, Oxford.

Bennett, R. J. (1977b), 'Consistent estimation of nonstationary parameters for small sample situations – A Monte Carlo study', *International Economic Review*, 18, 489-502.

Bennett, R. J. (1978), *Spatial time series: Analysis, forecasting and control*, Pion, London.

Bennett, R. J. and Chorley, R. J. (1978), *Environmental systems: Philosophy, analysis and control*, Methuen, London.

Bennett, R. J. and Tan, K. C. (1978), 'Allocation of the U. K. Rate Support Grant using the methods of optimal control', forthcoming.

Butkovskii, A. G. (1969), *Distributed control systems*, Elsevier, New York.

Caines, P. E. and Wall, K. D. (1972), *Theoretical foundations for methods of estimation and control in economic policy optimisation*, Discussion paper No. 3, Programme of research into econometric methods, London.

Chow, G. C. (1975), *Analysis and control of dynamic economic systems*, Wiley, New York.

Domanski, R. (1973), 'A general model of optimal growth in a system of regions', *Papers of the Regional Science Association*, 31, 73–82.

Garbade, K. (1976), 'On the existence and uniqueness of solutions to multi-period linear/quadratic optimal control problems', *International Economic Review*, 17, 719–731.

Hadley, G. (1962), *Linear Programming*, Addison-Wesley, Massachusetts.

Isard, W. (1975), *Introduction to regional science*, Prentice-Hall, New York.

Kalman, R. E. (1960), 'A new approach to linear filtering and prediction problems', *Transactions ASME, Journal of Basic Engineering*, D82, 35–45.

Klaassen, L. H. and Paelinck, J. H. P. (1974), *Integration of Socio-economic and physical planning*, Rotterdam University Press, Rotterdam.

Livesey, D. (1973), 'Can macro-economic policy problems ever be treated as quadratic regulator problems?', *Conference Publication 101*, Institution of Electrical Engineers, London.

Lowry, I. S. (1964), *A Model of metropolis*, Rand Corporation, Santa Monica.

Nijkamp, P. (1975), 'Spatial interdependencies and environmental effects', in A. Karlqvist, L. Lundqvist and F. Snickars (eds.), *Dynamic Allocation of Urban Space*, Saxon House, Farnborough.

Nijkamp, P. and Verhage, C. (1976), 'Cost benefit analysis and optimal control theory for environmental decisions: a case study of the Dollard estuary', in M. Chatterji and P. van Rompuy (eds.), *Environment, Regional Science and Interregional Modelling*, Springer, Berlin.

Pindyck, R. (1973), *Optimal planning for economic stabilisation*, North Holland, Amsterdam.

Polis, M., Goodson, R.E. and Wozny, M. J. (1973), 'On parameter identification for distributed systems using Galerkin's criterion', *Automatica*, 9, 53–64.

Pontryagin, L. S. (1962), *The mathematical theory of optimal processes*, Wiley-Interscience, New York.

Preston, A. J. (1977), 'Existence, uniqueness, and stability of linear optimal stabilisation policies' in J. D. Pitchford and S. J. Turnovsky (eds.), *Applications of control theory to economic analysis*, North Holland, Amsterdam.

Schoeffler, J. D. (1971), 'On-line multilevel systems', in D. A. Wismer (ed.), *Optimisation methods for large scale systems, with applications*, McGraw-Hill, New York.

Tan, K. C. (1978), 'Optimal control of linear econometric systems with linear equality constraints on the control variables', *International Economic Review*, 19, forthcoming.

Theil, H. (1966), *Applied Economic Forecasting*, North Holland, Amsterdam.

Turnovsky, S. J. (1977), 'Optimal control of linear systems with stochastic coefficients and additive disturbances', in J. D. Pitchford and S. J. Turnovsky (eds.), *Applications of control theory to economic analysis*, North Holland, Amsterdam.

Tustin, A. (1953), *The mechanism of economic systems*, Harvard University Press, Cambridge, Massachusetts.

Vishwakarma, K. P. (1974), *Macro-economic regulation*, Rotterdam University Press, Rotterdam.

Wilson, A. G. (1974), *Urban and regional models in geography and planning*, Wiley, Chichester.

Wolfe, P. (1959), 'The simplex method for quadratic programming', *Econometrica*, 27, 382–398.